DEAD TECH

A GUIDE
TO THE ARCHAEOLOGY
OF TOMORROW

PHOTOGRAPHS BY
MANFRED HAMM
TEXT BY
ROLF STEINBERG
INTRODUCTORY ESSAY BY
ROBERT JUNGK

HENNESSEY + INGALLS
SANTA MONICA
2000

DEAD TECH was originally published under the title
*Tote Technik – Ein Wegweiser zu den antiken Statten
von Morgen* by Nicolaische Verlagsbuchhandlung
Berlin, West Germany. Copyright ©1981 by
Nicolaische Verlagsbuchhandlung. English language
edition translated by Michael Stone. Copyright
©1982 by Nicolaische Verlagsbuchhandlung.
All rights reserved.

Reprinted with permission, 2000, by
Hennessey + Ingalls
1254 3rd Street Promenade
Santa Monica CA 90401

Printed in China

ISBN: 0-940512-22-X

Library of Congress Cataloging–in Publication Data

Hamm, Manfred, 1944–
    (Tote Technik. English)
    Dead tech : a guide to the archaeology of tomorrow / photographs by Manfred Hamm;
    text by Rolf Steinberg ; introductory essay by Robert Jungk.
        p. cm.
    Originally published: Sierra Club Books, c1982.
    Includes bibliographical references.
    ISBN 0-940512-22-X (paper)
    1. Industrial archaeology. 2. Technology – Social aspects. I. Steinberg, Rolf. II. Jungk,
    Robert, 1913- III. Title.

    T37. H3513 2000
    600 – dc21

                                                99-048091

CONTENTS

ROBERT JUNGK
THE RUINS COMPLEX

1

During the years when Hiroshima was being laboriously rebuilt, one of the few buildings not razed to the ground by the first atom bomb became the object of a heated quarrel. This was the former exhibition hall for industrial products, the bare dome of which had become a landmark amidst the desert of ruins and rubble. The majority of the citizens, many of them newcomers from other overcrowded towns, were of the opinion that the singular disaster which had struck the city on that 6th of August 1945 should best be forgotten as quickly as possible. The constant sight of that burnt-out hulk would not allow that. For that very reason the *hibakusha*, those leftover human wrecks, with death by atomic radiation close on their heels, wanted the stony witness to be preserved. The smoke-blackened hall, built in the years just before the First World War when people still believed in progress but were blind to its consequences, does indeed keep in mind the words chiseled into a gray tombstone for the almost 100,000 dead: "It must never happen again."

Though the defenders of this particular memorial won through in the end, they did not achieve their purpose. For the "atomic dome," as those old ruins are called by guides and sightseers from all over the world, does not act as a deterrent; it is merely picturesque. Its sight does not evoke ghastly memories but feeds the sentiments of the voyeur, providing those perverse thrills that will prompt thousands of people to satisfy their curiosity by crowding round any scene of disaster as they might buzz round the wreckage of a crashed plane.

In prewar photographs the building looks as unexciting, gray and dull as any other imitation Victorian or Wilhelminian architecture such as the Japanese Empire had favored, trying to match the imperial powers in the Far West. It was only the catastrophe that turned this humdrum piece of architecture into something unique, converting something ordinary into something quite extraordinary: an eloquent relic that does not echo the shouts for help and the agonized cries of the dying of that time.

For time does not only heal all wounds, it also blots out the memory of pain. Time renders the shocking worth telling; it transfigures despair and deceives us into believing that what has become meaningless can serve as a symbol. Ruins from the past, though they may be authentic enough, falsify history: in reality, they are almost always the dead bodies of past atrocities. The path of man is lined with desecrated temples, rifled burial chambers, razed castles and sacked palaces. We ought to recoil from them in horror as signs of violence, unscrupulousness, and crime. Instead, we turn to them reverently to admire what we should weep over.

2

This is not how we feel when looking at dead technology. These grotesquely twisted rods, burst casemates, rusty boilers, idle mine cages, stranded flying machines, and desolate spaceship ramps have not come down to us from some distant past; they still belong to our near present. They are not uplifting but rather ludicrous and horrifying at the same time. It must be a disconcerting experience for anybody who mistakenly thought such technical equipment to be precise and trustworthy, and relied on its functioning properly. The designers and producers are usually hesitant to speak about how vulnerable and prone to accidents their systems are. Yet what they prefer to keep silent about as long as possible is the true and trembling heart of any innovation.

To eliminate weaknesses, eradicate faults, reduce imperfections – these are the motive forces of research and development. We usually do not learn the whole truth about how fragile the creations of the inventors and engineers really are until they themselves have reached the next stage of development. With the knowledge that tomorrow they will be able to correct any mistakes made today or yesterday, they will criticize and belittle what they had just been rhapsodizing over.

To the man who puts his faith in progress, the relics of outdated technologies, preserved more or less by chance, are evidence of an interim stage of development when things are not yet working as they ought. Indeed, he will argue on the strength of comparisons with the obsolete how, in the course of time, everything is improving and growing ever more perfect. But if along this road he were to look back rather than ahead, he might in retrospect still be gripped by fear at the risks he had not been aware of at the time, the errors of which he might have become the victim, as despite all calculations they were not found out until much later.

The wrecks of smashed machines and engines, the closed down factories and abandoned laboratories and research stations – all bear the mark of Icarus, who had to crash to his death because his father, in a spirit of creative daring, had thought himself and his son capable of too much, too early. To be sure, the advocates of technological advances will offer a different interpretation. To them, the nightmares of previous failures will seem like the occasional whiff of belated panic on remembering the ordeal of an examination they had passed long ago. After all, they had succeeded then, confirming their belief that they will also succeed in the future.

In this way, after only a few embarrassing hours of the truth, the accident at the Harrisburg nuclear reactor, which happened against all expectations on the part of the nuclear technicians, was restyled a "stroke of luck," because, so the argument goes, it provided a splendid example from which to learn how to prevent something similar happening in the future. The former head of the International Atomic Authority went so far as to declare that such mishaps were needed to further the development of nuclear technologies.

3

At the turn of the second millennium most of us will think of less euphemistic interpretations and associations on seeing the ruins of modern engineering. We live for the most part in a man-made world, an increasingly ominous feature of which is the failure of some technical or mechanical equipment. For too many of our conversations seem to circle round some daily or even hourly breakdown. The car wouldn't start up, the plane couldn't take off, the train got in late. All the time some domestic appliance or other stops working: the fridge won't freeze up; the dishwasher's "gone mad"; on the TV set we get a crisscross pattern of lines; there's nothing on the tape of the recorder again; the neon tubes flicker; the telephone stays silent.

And in the world at large things are no different. The news is full of reports on accidents, breakdowns, catastrophes. Here an oil tanker has broken in half or a pipeline's burst; there we hear of a fire in an atomic power station, of containers exploding, of gases escaping; and even the well-tested safety technology of the chemical industry does not seem to be all that reliable anymore. Hardly a week passes without a cloud of poisonous gases escaping somewhere into the atmosphere, against all statistically substantiated probability. And how much more is happening on all those carefully guarded industrial sites? We know that only a fraction of the accidents at work are reported, but, even on the basis of such incomplete statistics, it was stated at a conference of the Stockholm Beijer-Institute in 1978 that in the United States more people had died in one year as a result of accidents than of cancer. Certainly, those responsible are trying desperately to check the disastrous consequences of technologies that are ever present, ever more efficacious, but never 100 percent efficient. They never wholly manage it. If they cure one defect, another shows up; they plug up a hole only to find a crack next to it. With those complex super constructions containing hundreds of subsidiary systems and thousands of interdependent parts, the failure of one small valve or the disconnection of a single

contact or the breakdown of an automatic control is enough to start off a chain of events – each wrong step in the process increasing the aberration of the next – that will turn "one of the miracles of modern technology" into a wreck.

The appropriate symbolic figure for such multiple technologies is no longer Prometheus, the dare-all, but Sisyphus, the eternal toiler. We should not allow ourselves to be misled about this by the successes of such highly remunerated and prestigious technologies as, for instance, space travel. They work so astonishingly well because many of their critical components are "redundant"; that is to say, they are built in as spares to take over when needed. In addition, such systems are supervised and rectified by an exceptionally large number of human controllers.

But even in the case of such pampered technologies, and despite the most stringent precautions, a tragic accident can happen. This is the lesson to be learned from the many thousands of pages of the protocol published by NASA on the accidental death of three astronauts. This model enquiry in the latest phase of the technological age makes one thing perfectly clear: even the most expensive, the most precise, and the most sophisticated machinery will fail if the people designing, building, or handling it make even the slightest mistake.

After all, what had led to those defenseless, high-flying idealists perishing miserably in the laboratory-tested capsule of their spacecraft was no more than some slight irregularities, some small omissions, and a little negligence. Since then the "human factor" is paid even more attention to than before as an important, maybe even the crucial, weak point in any "man-machine-system." Work psychologists are beginning to ask whether in this case we may not have reached our utmost limits, a point designers and engineers tend to disregard, their sole point of orientation being the efficacy of the machine. The reactor accident at Three Mile Island and the fact that there have been several operational lapses in the Pentagon's early warning system, which in times of crisis may decide the issue of war or peace, have added weight to such considerations.

It is becoming increasingly clear that man as the head on top of an all-powerful technological body is not yet, and probably never will be, able to control those forces that exceed his mental, psychic, and biological capabilities; it is a task that is almost certainly beyond him if he puts the huge power of his instruments at the service of wars and revolutions.

4

We are all haunted by visions that we are trying in vain to put out of our minds – visions of whole regions, continents, and even the earth itself being destroyed. The corpses of dead technology exhort us to be more prudent; they offer us predated evidence of the future; they are omens to warn us of the final collapse.

The towns bombed into waste shambles during the Second World War were prophecies we did not heed enough. On the contrary, industry, having forfeited the profits from armaments and war, saw in the mass destruction new opportunities for a profitable reconstruction. Indeed, the situation was reversed in an unusual manner. The people complaining were those whose property and equipment had escaped destruction, because now they had to compete with their old facilities against the most up-to-date. "We were really unlucky," the editor in chief of a large Cologne newspaper said to me, when he showed me through the undamaged publishing house soon after the war. "We have to carry on with our old presses now, as we won't get any means to modernize."

Our industrial society, constantly under pressure to apply the latest technological development, does not shy away from destruction. Quite the contrary, it is an implicit requirement. For new products to be sold, old ones must be got rid of. As the marketing strategists have discovered, this rhythmic cycle of destruction and renovation can be kept up without the periodic ravages of war, simply by implanting in the new products the seeds of their own rapid decay. Brand-new goods are already potential refuse when they are offered for sale. It is programmed with fair accuracy how soon they are to break down or to land – beyond repair – on a rubbish dump or in a scrapyard, ready for recycling.

In the case of wear-resistant materials, the policy is to ensure that the make or style of a product will date. With the media and advertising constantly pointing out to the public the alleged advantages of the very latest, the most modern, the consumer hanging on to what is old-fashioned is likely to gain the reputation of a crank or an old fogy who really belongs to where old bits and pieces are kept: in the lumber room. There, in its "proper place," yesterday's novelty might even acquire a certain curiosity value.

For in the objects we use, the houses we live in, the streets we walk through, even the shops that lure us with supplies of ever new products, is invested a piece of our own transitory existence, of personal memories, and it is not without regret that we see them disappear. That is why, together with our artificially stimulated hankering after the latest on the market, there is also growing a nostalgic longing for old or obsolete things. Single examples of former mass articles, saved perchance from the refuse collectors, are finding increasing appreciation. While they were available in great numbers, they were of no special interest to anybody. Now, however, as rare survivors they are highly valued and paid for accordingly, proving, wrongly in fact, that the period they originated from was not as transitory and culturally insignificant as might have been assumed.

If especially in Britain, the birthplace of the industrial revolution, such a thing as an "industrial archaeology" could come into being, and a struggle be waged for the preservation not only of gloomy old factories, derelict blast furnaces, and antiquated electric power stations, but even of those terrible sweatshops, the textile mills of old that were so much hated in their time, and also of what is left of the old slums, we may take it as an indication that people regret the way change has

been speeded in our days. We bewail the rush of our existence and lament the early death of an era transfigured in our memory. Actually, the rubbish dumps on the outskirts of our cities, composed of half-used waste, broken pieces of furniture, burnt-out bits of machinery, and smashed-up crockery, are far more typical of the throwaway society of the twentieth century than the odd shards kept as souvenirs that by accident or neglect escaped the mass graves of the industrial age.

Those mountains of refuse, of repellent formlessness, filling our nostrils with the stench of decay, show up the lack of consideration and of reverence of a civilization which, like a predatory animal, uses its huge mechanical tools of mastication to crush, digest, and discharge again in no time what has taken centuries and even millenniums to grow.

In a few decades, only half of all the woods of our globe will still be standing. Each year hundreds of unique species of plants and animals become the prey of this devouring civilization, sacrificed irretrievably and forever on the altar of its gluttonous greed. Even the creatures in the depth of the sea are no longer safe from outside interference. Remote-controlled scrapers are digging themselves through the sediments that have taken millions of years to sink down into the depth which no light can reach. They come up with cold and dead metal, sacrificing to this end the ever self-renewing life cycle of the maritime biotope.

Why should this race show more respect for man-made things than for what nature has created? Avarice and the drive to be active at all costs have thus been detrimental, too, to the artifacts. The great wastrels accelerate the inevitable entropy – chaos – in which all matter will dissolve itself one day. The living have become extremely proficient helpmates of death. Never before could he boast of such excellent accomplices.

5

At the beginning of the 198th decade of this second millennium, I once heard a prophecy out of the mouth of a dark-skinned man that has haunted me ever since. He was a descendant of the Australian aborigines and had traveled to Europe with some of his friends, to protest in Bonn against West German firms participating in the mining of uranium in the deserts of the fifth continent. There are rich deposits of the metal needed for the production of atomic energy in those out-of-the-way regions where the natives have their secret places of worship. These holy hills have now been dug open by bulldozers and drilled through by powerful drills. He described the sacrilege while we were driving through the densely built-up industrial landscape of West Germany in a small VW bus. When he had finished we all fell silent, looking out onto the motorway and beyond, where through the smog-filled air we could make out skyscrapers, blocks of flats, and workshops.

Suddenly one of them spoke out: "All this here is still standing. And all the cars and people are still moving. But I know what it will be like one day. I see an empty, polluted plain with nothing growing on it, where nothing will be growing and no one will be living for centuries.

For the missiles filled with the ore from the desecrated living places of our gods will fall upon those who did this or allowed it to happen. They themselves produce the tools with which vengeance will be wreaked on them, and they are not even aware of it."

It is an entirely new form of disaster that is threatening mankind: the regions struck by atomic catastrophes will not only be annihilated, they will remain uninhabitable for a long time. They will become waste areas that no one will be able to visit with impunity (unless he were to put on a clumsy protective outfit): they will be radiation ruins!

All previous ravages had grass growing over them, and the loss of life was made up within one generation by new births. Never before in history have the effects of war or of catastrophes been irreversible. Each blunder was something one could learn from, every error one could afterward try to repair. This time it will be different. For the latest energy and military technology reaches far beyond the time of its application, affecting the fate of future generations. It will survive both itself and its creators for years to come. The radiation set free – unintentionally in times of peace, deliberately in a war – will go on being active for tens and hundreds of years, in some cases even for thousands and hundreds of thousands of years, producing masses of human wrecks – at first people who are physically disabled and subsequently, in all probability, genetic freaks. From a dead technology to a technology that over generations remains active as a bringer of death by heat and poisons – what a step, what a fall forward!

As long as these iron angels of death remain among us, no one can look forward hopefully to what's to come. There are not only the visible reminders of dead engineering and dead technologies, we are also haunted by visions of a grievously damaged future world. These shadows lie ever more heavily upon our hectic existence, with its whirl of sensations, its successes for a day, its quick careers. They impress on our minds that we must be on the wrong track, that the end of our future has begun already, and that history is racing toward a tragic climax if we do not alter its course very soon.

6

If we look for the reasons why today hardly anybody dares to plan ahead, why in our era no buildings are erected to challenge time, why no great epic poems are written, why no comprehensive philosophical systems are devised to order and interpret the thousandfold torn web of knowledge – if we ask why there is so much superficiality, irresponsibility, and shortness of breath, and why so many young people escape into the dream world of drugs – the answer is that the common denominator for the prevalent resignation, indifference, and boundless despair lies in our lack of faith in the future.

All those who no longer dare to believe in a brighter world of tomorrow, because there are so many indications against it, are victims of a neurosis which I would like to call the "ruins complex." This factually and rationally substantiated attitude becomes neurotic in character once it surrenders to the idea of the inevitability of destruction as the only determinant and supposedly inexplicable power.

It is quite true, though, if we project present trends in a straight line into the future, that we are heading for extinction. But it is a peculiarity of man that he can respond to a challenge by radically changing his way of thinking and acting. For instance, he can try to invent in his head a different kind of future. Why should it be beyond him, instead of a doomed and deadly technology, to conceive and try out one that is living and life-giving?

Just as there are different styles in architecture, there must also be different styles of engineering and technology. If our current way of dealing with human beings and nature is predominantly coarse and brutal, another way is quite as conceivable: gentle and considerate, no longer working against our biosphere but within it and in keeping with it. Such a gentle technology is being tried out in many places already. This is still in an initial stage, but a growing number of heads and hands are engaged in efforts to bring about a higher development of our technological means.

Dead technology stands for everything that frightens all of us. Are we to give in, are you going to give in, shall I give in to this writing on the wall? It is very tempting, for to give in is always easier in the beginning than to resist. We must not tire in our efforts to draw an impressive picture of the apocalypse threatening us, nor to speak of it all the time so that, seeing how little time is left to us, we can quickly mobilize as many people as possible against it.

Albert Camus, on receiving the Nobel Prize, spoke as follows of those who in the face of an almost lost war were not prepared to give up hope: "They were compelled to forge for themselves an art of living in calamitous times, in order to be born a second time and to fight with an open visor against the instinctive longing for death that seems to be at work in our history".

Since the time that speech was delivered in Stockholm more than a quarter of a century has passed, and the number of those trying to bring about a change of mind has grown quite considerably everywhere. This is comparable to the rise of Christianity at the end of what today we call antiquity.

The whole world need not be left in ruins – a dead planet is not inevitably the consequence of the sum total of blunders – but it means the bankruptcy of an attitude of mind in which people blindly and arrogantly thought that there were no limits to what they could do and control. That is, indeed, what will be left: the debris of such an ideology.

THE RUINS OF WAR
VERDUN, THE MAGINOT LINE, THE ATLANTIC WALL

From the walls of Jericho to the armored cupolas of the Maginot Line, from the rocks of Masada to the casemates of Verdun: the list is long of the historical monuments of fortresses once thought impregnable, or that were meant to appear so, until they proved otherwise. Their deserted ruins can be traced throughout the world's history, like geological deposits of war. The Great Wall of China, the greatest single technical construction ever undertaken by man, can even be made out from the moon with the naked eye.

During the Stone Age, the first communal buildings were devoted to defense or to the cult of the dead. In troubled times, people and armies would seek safety behind walls, gates, and towers. They would entrench themselves behind fortifications made of pounded earth, stones, bricks made of clay, concrete, and plated steel – sometimes well above ground, sometimes well below. As on a revolving stage, those under siege yesterday themselves lay siege the day after – and vice versa.

Jericho was a fortified town thousands of years before its biblical walls miraculously collapsed under the trumpet blows of the priests and the shouts of the children of Israel. Excavations on the Hill of Ruins, at Tel Es-Sultan in the Jordan Valley, have brought to light the oldest known fortifications: a 16-foot-high rampart with a peel tower to be ascended from inside of about the same height and 30 feet in diameter. According to the archaeologists, these fortifications date back to the preceramic neolithic age, i.e., a time when the burning of bricks and of vessels of clay was not yet known.

When the Israelites reached the Promised Land to the west of the river Jordan about 1300 B.C., Jericho's oldest fortifications had already been sleeping for a long time under piles of sediment. As told in the Book of Joshua, spies were sent out to study the lay of the land from a military point of view and to find out whether the Canaanites were living in "tents or in fortified places." They did live in fortified places. Jericho was reconnoitered with the aid of Rahab, the harlot, who hid Joshua's spies in her house at the town wall, for which reason the sword of Israel spared the traitress.

As, according to the Old Testament, in the Holy Land, so in antiquity almost all capital cities and trade centers were "walled up to the sky." Nimrud, Nineveh, Babylon, Jerusalem, Athens, Carthage: How many of the seats of power in ancient times did not boast of indestructible fortifications? Greek legend has it that those who built the citadel of Mycenae must have belonged to the race of the one-eyed giants on account of the mighty square stones piled yard-high one upon the other. The neolithic cyclopean walls withstood any frontal attacks. Time and again beleaguering armies tried to starve or to trick fortified towns into submission. For ten years, the war swayed back and forth outside the gates of Troy until the Trojans fell for Ulysses' ruse with the wooden horse. Ignoring Cassandra's warnings, the Trojans themselves admitted their enemies to the city in the belly of the monstrous animal.

On reliefs dating back to the ninth century B.C., the Assyrians can be seen employing the first battering engines: mobile battering rams with or without protective cover, known simply as "rams" or "tortoises." The beginnings of an arms technology are marked by multiple-storied assault towers on wheels, missile engines and catapults of ever greater caliber and range. The rapid development of mechanized warfare inspired the military writers of antiquity to theoretical works on the siege and defense of fortified places, just as the German Heinz Guderian or the Frenchman Charles de Gaulle would tackle the Blitzkrieg in the age of motorized tanks and other armored vehicles.

One of the largest war machines of antiquity was set in motion against Rhodes in 305 B.C., by Demetrios, the "wrecker of cities." Thirty-four hundred soldiers had to push the nine-storied siege tower, which reached a height of between 100 and 148 feet. But the defenders managed to halt the monster and to seize it in the end. From the sale of the materials the town financed a statue to give thanks to Helios, the sun god. The bronze-plated Colossus at the entrance of the harbor crashed to the ground eighty years later in consequence of an earthquake in the Aegean Sea, but the wreckage left by what had been one of the seven ancient wonders of the world was still lying around for another 900 years.

To bring their heavy armaments into position, both the Greeks and the Romans sometimes had to undertake quite frantic feats of civil engineering. Tyros, prosperous Phoenician port in the eastern Mediterranean, resisted Babylon's warriors for thirteen years. Two centuries later, Alexander the Great sealed the fate of the island fort by the construction of a dike from the mainland for his assault troops. The stony banks built in 332 B.C. can still be made out from the air because of the broad drifts of sand.

The Roman general Flavius Silva drove his siege ramp right up to the rocky plateau of Masada. That's where the last fighters of the Jewish uprising of 66–73 A.D. offered desperate resistance, 1300 feet above the Dead Sea. In a hopeless situation, they heard the heavy pile driver gradually break up their bastion. Knowing that they had to expect from the victors nothing but slaughter, torture, or slavery, the rebels chose death for themselves and their families. The Israeli ex-general Yigeal Yadin excavated the fortress of Herodes in the 1960s, his findings confirming with surprising accuracy the description given by the contemporary historian Flavius Josephus of the heroic last stand of the Jews at Masada.

Against the bellicose Teutons and Celts, the Roman Caesars used to fortify the northern borders of their empire. In form of the Limes, the Roman legions left a rich source of archaeological discoveries behind in old Germania. The continuous defense line stretching from the Rhine to the Danube consisted of stockades, watchtowers, signaling installations, and citadels for the troops of occupation. In Britain the Romans closed off their province Britannica from Scotland by building Hadrian's Wall. Like a gigantic caterpillar, it is still snaking its way along the waist of England, 78 miles across Northumbria.

Steel cupolas looking from afar like cheese dishes or flying saucers are gazing from the undulating hills of Alsace-Lorraine into the valley of the Moselle. In wet weather they shine like newly dug molehills, and, indeed, they had a similar function.

They are the superstructures of mammoth bunkers spreading out underneath fields and meadows like moles's burrows. "Europe's largest and most complex architectural and military folly" – so called in the expert study *Walls of War* – is for the most part hidden underground.

The invention of gunpowder and cannon balls, the development of grenades, mine throwers, high-trajectory guns, and mortars represented decisive tactical and technological stages leading to the construction of fortified cities underground. Between the two world wars, the armored works of the Maginot Line with their retractable gun turrets were the most modern such fortifications in the world.

These subterranean chambers, sunk several floors deep underground, were thought to be immune to even the heaviest artillery fire and aerial bombardments because of the yard-thick ceilings made of reinforced concrete. The extensive system of connecting tunnels and gangways covered 93 miles, almost as long as the Paris Metro. This interlinked network of forts, observation towers, antitank ditches, gun emplacements and casemates extended from Strasbourg to Montmédy. The backbone of this zone of fortifications, reaching in places a depth of 24 miles, consisted of thirty-nine large-scale forts, known as *ouvrages*. In respect of their total firepower and technical equipment, they stood comparison with dug-in ships of the line. The superfort Hackenberg, for instance, had a crew of 1100 men, and its artillery could shell the enemy at a rate of four tons of ammunition per minute. Its power station could have supplied a town of 10,000 inhabitants with electricity.

André Maginot, minister of war, had demanded 3 billion francs for the building of the line of defense that was to bear his name in 1929. Before the National Assembly, he conjured up the danger of new aggressions from the East: it would be better to stop the invaders by a wall of concrete, concrete being cheaper than human lives. This was the trauma of Verdun – the slaughterhouse of a whole generation of men during World War I. The illustrated guide of the local Éditions Lorraines put it with greater solemnity: "Approximately 800,000 soldiers died a hero's death on the battlefields around Verdun."

On the way to the cemeteries and memorial places of 1914–18, along National Route 403, we reach the armored fortress Douaumont and the Fort de Vaux. Situated upon a ragged plateau on the right bank of the River Maas, they constituted a third line of defense for Verdun toward the northeast when, on the 21st of February 1916, a seven-hour barrage from over a thousand guns opened up. The German chief of staff, Erich von Falkenhayn, with a large-scale attack on Verdun, unleashed the greatest battle of annihilation in history. No less than 8000 heavy shells per day rained on Fort de Vaux and the surrounding countryside, for weeks on end. Yet it took the German fusiliers until June 8 to occupy the casemates and vaults of the fort. The offensive with which Falkenhayn had hoped to crush the French army before Verdun, sapped away the strength of the Germans like a festering wound. When in August 1916 the murderous undertaking was called off, they had lost 337,000 men, dead, wounded, or taken prisoner. The French, in turn, shelled and attacked Douaumont in vain. The then strongest armored fort in the world had withstood even being fired at by Krupp's "Dicke Bertha", the 16 inch giant mortar. Yet in 1915, Marshal Joffre had withdrawn his troops from the fortress, which right at the beginning of the battle of Verdun had fallen into the hands of the Germans in a surprise attack. The recapture of this cornerstone by the French in December 1916 crowned their victory at Verdun. It had cost the nation a terrible bloodletting in husbands, fathers, and sons, as the long lists on the war memorials everywhere in the country bear witness.

In World War I, the losses of the French were proportionally greater than those of the Germans, with about 1,5 million dead – or approximately 10.5 percent of the men old enough to serve in the forces – buried on the battlefields of Flanders and the Champagne. The industrial centers in the north lay in waste; the finances of the state were ruined. It made no difference that France regained Alsace-Lorraine and occupied the Ruhr. No wonder that the idea suggested itself to the weakened victorious power of 1918 for the first time to put an unassailable wall between itself and the restless Germans.

Marshal Pétain, the victor of Verdun, raised the idea of a fortified line to an indisputable doctrine of defense. In 1922 he demanded the erection of a "French Wall" along the eastern frontier. In the late summer of 1933, Minister President Daladier could report the fulfillment of that demand. After an inspection of the Maginot Line, he announced outside the town hall of Sarreguemines that, after four years of superhuman efforts, France was safe from any surprises. However, there was a gap in the line of fortifications on the border between Belgium and Luxembourg. The French general staff pooh-poohed the idea of a massive frontal attack by tanks through the woody gorges of the Ardennes. Yet this was precisely the way the German offensive proceeded on the 10th of May 1940: through the hole at Sedan toward the River Maas. Meanwhile, the defense forces were waiting in vain for the enemy at the vision slits and periscopes further south. "On ne passe pas!" (No one shall pass here!) it said boldly on the large forts. And, indeed, no one tried to, apart from some local breakthrough skirmishes. The Battle of France was not waged in the forefront but in the rear of the fire-spitting dragon that was the Maginot Line.

After the armistice, the forward defenses of the Third Republic fell into the hands of the Germans almost intact. In contrast to Verdun, the sorry remnants of a strategic miscalculation do not lend themselves to be turned into a memorial place. A breath of futility, the melancholy of bewitched palaces, is drifting over the solitary casemates in the autumn: art deco in concrete – wasting away behind ivy, brushwood, and dry rot.

The national administration of state property has managed to sell some of the bunkers to private connoisseurs and estate agents. But the *grands ouvrages* are not for sale. As fateful landmarks of a frontier region, they play a similar role in the local traditionalism as do the old pitheads and furnaces in the Lorraine Basin. The Maginot Line has become a symbol of local patriotism. People predominantly from this region had built and guarded it in the 1930's. Nor is there any reason to feel ashamed because of its failure in 1940. After all, under Charles de Gaulle's Cross of Lorraine, France did finally emerge as one of the victors in 1945.

"Visitez la Ligne Maginot," says a travel brochure of the town Thionville, recommending for an inspection not just one, but three "representative fortified works." The municipal council is using the military remains as a tourist attraction. Old soldiers' associations and traditional societies keep the antiquated fort technology in working order. Regularly in the spring, the generators are checked, the guns polished, and the casemates aired. Through the mole tunnels of Fort Hackenberg, deep underground below Veckring, is rattling the electric narrow-gauge railway. When autumn comes it is all over; there are not enough visitors for such attractions; the concrete dinosaurs lapse into hibernation.

Just a few hours away by car, at the Pas de Calais, one can find the relics of the Atlantic Wall of 1942–44. If Hitler had been able to finish building his "Fortress Europe" as planned, it would have been longer than the Chinese Wall.

In March 1942 – the Germans having lost the initiative in the West – the commander in chief of the Greater German Reich issued Order 40, "Coastal Defense." To repel any Anglo-American landings, he, too, could think of nothing better than the rigid defense strategy of a Maginot Line. He announced the coasts of the occupied continent, from the North Cape to the Bay of Biscay, to be the main line of battle, behind which there was to be no retreat. Racing against time, he began in the same year with the construction of fortifications, the French coast being the key area. That's where the *Organisation Todt* mobilized a labor force of about 291,000, mainly conscripted workers, for the project "Atlantic Wall." Albert Speer, then minister of armaments, reports in his memoirs that Hitler himself planned the main points and the succession of these fortifications right down to the last detail. With reference to his own experiences as a front-line soldier, the amateur architect even drafted sketches of various types of bunkers himself.

Hitler wanted to secure his western flank against the Anglo-American Second Front through a linear coastal defense of 15,000 concrete forts and dugouts from Holland to the Spanish border. The Atlantic Wall was as yet incomplete when, on the 6th of June 1944, "the longest day" dawned on the invasion beaches of Normandy. Up to that time, Speer had squandered 18.9 million cubic yards of concrete at a cost of 3.7 billion marks, as well as 1.2 million tons of iron, on a propaganda fortress that on D-Day, the day of the actual landing, still had no roof to cover it. In the face of the fact that the Allies had absolute control of the air, the Atlantic Wall, with its motley collection of second-class captured guns, seemed as much out of date as the Maginot Line had been during the Blitzkrieg of 1940. Even the gap of Sedan had not been forgotten by the German boffins. While the fortifications along the English Channel were in an advanced state of readiness, a weak point was left along the Calvados coast. The culprits this time were German naval experts who had declared a landing in Normandy between the mouths of the rivers Orne and Vire as highly improbable because of the underwater reefs and other handicaps.

After the liberation of France in 1944–45, many of the deserted bunkers and their ammunition were blown up, an act of revenge against objects. The chunks of concrete were still identified with the occupants themselves, just as though they had left their steel helmets on the beach. Some of the fortifications were turned to communal or private use, others were consecrated as memorials or were just left standing there. One way or another, France could not get rid of the unloved range of bunkers. Overgrown by blackberry bushes, the blocks of concrete fit into the landscape of fields and meadows like erratic hillocks. Along the Channel coast and elsewhere, the dilapidated piles of ruins are sinking by inches into the ground under their own weight: another Carthage in the vicinity of Calais, the ruins of which, according to one's fancy, are either half buried of half dug up again.

Like prehistoric barrows, the anonymous concrete shells of the Atlantic Wall appeared to the Parisian architect and town planner Paul Virilio to have a religious function. In his book *Bunker Archéologie* he describes them as altars and catacombs of an archaic creed of fortifications in which are joined myth and propaganda. "A whole series of cultural reminiscences passed through my mind: the mastabas, Etruscan tombs, Aztec structures … as if this light gun casemate had something to do with funeral rites and as if the *Organisation Todt* had finally done nothing else but create a religious space …."

Fort de Vaux/Verdun

Fort de Vaux/Verdun

Artillery Block, Rochonvillers/Maginot Line

L'Ouvrage Hackenberg/Maginot Line

Signpost/Maginot Line

Gun Cupola/Maginot Line

Artillery Block/Maginot Line

Gun Cupola and Lookout/Maginot Line

Batterie de Longues/Atlant c Wall

Bray Dunes/Atlantic Wall

Bray Dunes/Atlantic Wall

Bray Dunes/Atlantic Wall

Bray Dunes/Atlantic Wall

Bray Dunes/Atlantic Wall

RUST IN PEACE
THE GRAVEYARD OF STEAM AT BARRY DOCKS

Tears are rolling down the steam locomotive's moon face. A railway fan had painted it on her fire door, adding a cry for help: "Please don't let me die." We are in South Wales, in the Graveyard of Steam at Barry Docks.

It was here, in the county of Glamorgan in 1804, that the first steam train ever was rolling over iron rails. It was a peculiar-looking vehicle with an oversize flywheel, bare rods protruding, and various sized cogs for the power transmission. The whole thing took place on a 10-mile-long track belonging to a pit owner by the name of Hill, leading from the ironworks at Pen-y-darran to Cardiff. An immigrant Cornishman, Richard Trevithick, working as an engineer at Samuel Homfrey's ironworks, wanted to use the track for a trial run of the high-pressure steam locomotive he had built. He claimed it would shift at one go a load of ten metric tons, which was many times the load managed by draft horses.

Hill dismissed this as plainly foolish, staking £500 sterling against it. Buth with five wagons filled with iron ore and, in addition, seventy passengers, Trevithick's "tram wagon" hauled no less than 25.4 metric tons all the way. It took the clumsy "iron horse" all of four hours to cover the distance, but there could be absolutely no doubt about its superiority in strength as compared with the "oat-driven motors." However, Trevithick was not yet able to give sufficient proof of his hauling engine's economical advantages, due to the rails (used only to horse-drawn cars) giving way under the excessive load at the next attempt. So the great-grandmother of the steam locomotive fell into oblivion again.

About 150 years after the tram wagon had made its debut, Britain's last steam locomotive was baptized Evening Star at a launching ceremony in the Swindon Works: a rail giant weighing approximately ninety tons with a haulage capacity of 18,000 tons. It belonged to the most modern class of locomotives, 9F, introduced by British Rails in 1954 mainly for the haulage of heavy goods. Just a few years later British Rails opened up its diesel program; in the face of cheap oil, even the most up-to-date, pulverized-coal-fire technology could not compete with the internal combustion engine. One after the other of the hissing and snorting steam locomotives vanished from the public rail net, bringing to an end a glorious era of railway history marked by many popular landmarks of the technological age: railway stations like cathedrals, urban rail proliferations, water towers, and signal boxes. In the mother country of the industrial revolution, the last of about 40,000 steam locomotives, the Oliver Cromwell, an express locomotive in the streamline form of the Britannia-class, was taken out of service in August 1968. The Evening Star had landed in the historical railway collection at York in 1965, already as a national relic, after only five service years.

Most of the locomotives taken off the line ended in a scrapyard without so much as a thank you. However, in the county of Glamorgan, where Trevithick had built his memorable tram wagon, a track section gradually filled up with the cooled pride and glory of the British steam railway from 1957 onward. As though they had secretly arranged a demonstration of protest against their undignified dismissal, 368 veteran locomotives bearing the emblem of British Rails, a lion rampant, had gathered finally in the scrapyard of Woodham Brothers.

The open-air mausoleum at Barry Docks, near Cardiff, has entered railway literature already. In 1974, Brian Handley immortalized it in a book of photographs to which he gave the title Graveyard of Steam, and that was well done. For the days of the Graveyard of Steam are numbered. From all parts of England, railway fans have been and still are flocking to a scrapyard where Britain's last coal-fired locomotives are corroding by the seaside.

According to Mr. Woodham's estimate, the number of people paying a farewell visit is approaching the 2-million mark already. On weekends, they can be seen wandering with devout attention over the abandoned siding. "It is sad to think how they once puffed and thundered across the country," one is murmuring to himself.

Young train spotters, armed with cameras and notebooks, are out on a reconnaissance patrol. With expert eyes, they fix the year of construction and class of all types gathered there buffer to buffer and funnel after funnel. The railway amateur may recognize the distinguishing marks of the various designs, but these train achaeologists know the exact origin of each of the antique monsters. Barry Docks constitutes a unique place of discovery: there are still more than 100 locomotives of various age groups thronging the sidings.

Among them sits the beefy Churchward, a venerable construction from the beginning of 1900, which had been toiling away in the local passenger service of the western region until she was shunted out in 1965. You'll find here wasting away the last standard-class locomotives of 1951–54, known by their serial numbers consisting of five digits beginning with the figures 7–9. Also, the former celebrities of the King and Jubilee classes: purebred express locomotives from the 1930s, which together with their luxury trains have since been transfigured into subjects of legend.

The Midland Action Group has got busy on the express locomotive No. 6023, King Edward II. They have removed the rust from the superstructure of the majestic steam giant, painted the boiler green, and lubricated the connecting rods. Britain's railway enthusiasts did not remain inactive in the face of the threatening steam massacre. They have collected more than 100 of the inmates of the Graveyard of Steam from Barry Docks for restoration purposes already. This rescue work is continuing. Almost every third or fourth steam locomotive bears a notice from one of the so-called preservation groups, announcing similar intentions. "Sold," "Reserved for," or "Do not remove parts," it says on the iron bodies. To protect her from further decay, the new owners of one such locomotive covered her with a tarpaulin. A box for messages and a collection box is attached to No. 7093. The 7093 Fund is aiming to restore the Foremake Hall, built in 1949, to her former glory and so save the last example of her class from the blowtorch. In addition to a lot of enthusiasm, such an undertaking requires generous donations. To buy one of the rusty locomotives you'd need between

£2,000 and £10,000, according to the scrap value, plus the costs of transport by low loader and of the materials necessary for the restoration. By the time the happy end is reached, the expenses might easily add up to £50,000 or more.

Time is getting short. In two to three years the reprieve for the remaining steam invalids will be running out; the oldest among them are corroded beyond salvation already. The continuous salty breeze from the Bristol Channel has rotted away their superstructures; underneath the boilers' curves one can see the asbestos layer shimmering like putrid flesh. The rust brown shells have long since been reclaimed by nature. Spiders have invaded the drivers' cabins; birds are nesting there. In the course of years, some of the tenders have turned into slimy rain tubs from which wild plants and even trees are sprouting.

Some of the iron horses display on the walls of their boilers cries of help ("a good friend") or bitter charges ("you killed me"). Indeed, the way they are standing there rotting away, their naked bodies exposed to all weathers, they look like some mythical creatures of the machine age. As a matey farewell greeting, somebody has written in white oil paint "Farting Flyer" on the belly of a dying express locomotive, and another penned on the base frame of a goods engine, "let it rust in peace."

"Graveyard of Steam," Barry Docks/South Wales

"Graveyard of Steam," Barry Docks/South Wales

"Graveyard of Steam," Barry Docks/South Wales

"Graveyard of Steam," Barry Docks/South Wales

"Graveyard of Steam," Barry Docks/South Wales

"Graveyard of Steam," Barry Docks / South Wales

"Graveyard of Steam," Barry Docks/South Wales

Anhalter Goods Station, West Berlin

SUNKEN HARBORS
NEW YORK, BRIGHTON, PORT WINSTON

The Mediterranean, the cradle of seafaring, knows many sunken harbors.

Some were destroyed by war or the forces of nature; others were left to decay or were buried underneath the wash. Tyros and Carthage, once mighty maritime forts of the Phoenicians; Ostia, at one time the naval port and corn harbor of the Roman Empire; the port of Herodes at Caesarea: these names stand for other seafaring centers in antiquity that lie buried around the Mediterranean. Archaeological remains point to harbor installations of considerable size and facilities.

A huge lighthouse, approximately 430 feet in height, was counted as one of the Seven Wonders of the World in ancient times. The Greek master builder Sostratos had erected it in the third century B.C. at the entrance to the harbor of Alexandria, dedicating it "in the name of all seafaring men to the divine rescuers." From the rock island of the same name the famous Pharos loomed on the horizon for almost 1,500 years. Contemporary travel reports contained enthusiastic descriptions of the edifice, which to the Moorish prince and Mediterranean traveler Idrisi seemed at night „like a glittering star."

In the twelfth century, the Arabic conquerors of Egypt extinguished the beacon of light. From then on the tower served as a mosque. An earthquake caused its collapse in 1375. The ruins could still be seen in the fifteenth century, as J. Sprague de Camp tells us in his book *The Ancient Engineers.* All one can see there today are a few rocks washed by the sea.

*

A harbor front waiting for demolition extends on Manhattan's west side from Fifty-ninth Street right down to the Chelsea Docks. Swedish-American, United States Lines, Cunard: those names on the deserted piers seem like obituaries upon the death of a century of ocean liners. Yet hardly twenty years have passed since about 900,000 ships' passengers entered or left the New World here on the banks of the Hudson. One ocean liner after another used to dock in front of Manhattan's skyline, just as the jumbo-jets do today along the gates of New York's John F. Kennedy International Airport. After the introduction of the Clipper had reduced a transatlantic flight to seven or eight hours nonstop, the traveling public deserted the shipping lines and turned to the airlines. Even the best-known and fastest luxury liners were not able to keep up the competition against the cheaper, long-distance jets. The British Cunard Steamship Company, the first shipping line to have opened up a transatlantic mail service in 1840, laid up its flagship, the 81,000-ton *Queen Mary,* in 1967, after her 1,001st Atlantic crossing. The *United States,* since 1952 holder of the Blue Ribbon, gave up two years later, even though as the world's fastest turbine steamship she had cut down the time of crossing from Europe to New York to less than four days or, to be more precise, to three days, ten hours, and forty minutes.

In 1973, 25,000 passengers crossed the Atlantic by ship as against 12 million by air. At the Superliner Terminal between Forty-eighth and Fifty-fifth streets in the heart of Manhattan, where formerly sometimes three or even four queens of the ocean had made fast, an occasional cruise ship may dock today – in the summer. As a passenger port New York has lost out to Port Everglades in Florida. The abandoned reception halls and goods sheds now serve as garages or exhibition halls, if they don't dilapidate or rot away, with trees growing on their roofs. New York City opened the dead harbor area along the Hudson for nonmaritime purposes in 1970.

Even the traditional ferry and freight service on the west side has come to a stop. The two railway piers, 100 and 101, behind Fifty-ninth Street are burnt down; their iron skeletons are slowly sinking into the water. The Chelsea Docks further down were made redundant by the container transshipment now concentrated mainly on Staten Island and Brooklyn.

The introduction of automation at sea put other goods harbors out of work. Liverpool, known in its heyday as the port of a thousand ships, is but a shadow of its former self. In London's dockland along the bends of the river Thames, where once the heart of the British Empire was beating, the cranes stand idle. The traditional docks and wharves in Wapping, Rotherhithe, and Woolwich are about to disappear from sight. Container ships and freighters carrying bulk cargoes dock outside London at the new Tilbury Docks near the mouth of the river. The computers are in command now.

*

England's loveliest pleasure ruin reaches 1,000 feet out into the English Channel on piles. At the head end it carries a games and theater pavilion, which, with its turrets and arcades, looks in the morning haze like an oriental palace floating over the water. But the red letters on the roof look dull – the majority of the lightbulbs have dropped out of their sockets.

Because of its dilapidated state, Brighton's West Pier was closed at the end of September 1975. The portal facing the embankment has been nailed up; the roofs of the pay-boxes have been covered with barbed wire to scare off anyone wishing to climb up. Will the West Pier be restored and made into a tourist attraction with a fairground, as is the intention of a building contractor? Will it be demolished at the expense of the taxpayer, or will it collapse by itself one day? Red warning signs tell us that it is a Dangerous Structure, this gem of Victorian maritime architecture, now waiting for better times or the end.

In Queen Victoria's days, it had become almost a national mania to build piers on graceful piles far out into the sea. Finally, at the turn of the century, eighty-five such pile structures on wooden or cast-iron supports adorned the coast of Britain. It was time to draw a new map of England in the shape of a huge porcupine, was the mocking comment of a gossip columnist on the hedgehog look of the British Isles.

Most of the piers – the longest at Southend-on-Sea has a length of 7,000 feet – were not just simple landing stages but typical landmarks of Merry Old England: pleasure piers that beyond any nautical purposes served as recreational facilities and public entertainments. With their gangways, concert platforms, promenade decks, and pavilions, they resembled luxury liners that never left their moorings. Aspiring and fashionable resorts like Blackpool, Scarborough, Margate, or Hastings felt the need to adorn themselves with a pier or two, to flatter the elegance of their visitors and even to suggest a touch of frivolity. To this end they secured for themselves the engineering skills of Eugenius Birch (1818–84), the most eminent builder of piers of his age.

Brighton, the "Queen of the South Coast," indulged herself with a masterpiece that "set the standard for a generation of piers to follow," according to Simon H. Adamson, who wrote an expert history on the subject (Seaside Piers). Originally, Birch's design provided for four symmetrical kiosks on the landing promenade. Later additions such as the games pavilion and the music hall changed the silhouette of the West Pier without, however, spoiling its grace and harmony. Emperor Napoleon III landed here in 1870, praising the pier as "Britain's finest structure."

Meanwhile, the piers are beginning to suffer the fate of old age. Debility, war and storm damages, fires, and collisions have decimated their ranks, so that no more than twenty-five veterans are likely to survive until the year 2000. Brighton's West Pier ("the best pier") is not the only one with its corroding superstructures wasting away in uncertainty. However, a local citizens' initiative and the Victorian Society are prepared to man the barricades against its demolition. To this decorative iron construction from the great days of British naval power applies in particular what the Observer wrote about those once so popular pleasure piers: "They are," the paper wrote, "in their way as irreplaceable as medieval cathedrals."

*

A legacy of the greatest amphibian landing operation in the world's history is sinking into the sea off the steep cliffs of Normandy. In all works dealing with the history of World War II, it is mentioned by its code name "Mulberry." Mulberry was an artificial supply harbor that Britain, being a naval power, had built specially for the operation known as "Overlord"; it was towed to the invasion beaches in the bay formed by the mouth of the river Seine, from which, on 6 June 1944, the Allies started their second front on Hitler's Fortress Europe with two armies, 6,697 ships, and 14,600 airplanes.

Greater Germany's supreme commander was tricked into believing that the invasion was a feint, because there were no harbors in the area where the first bridgeheads were established. In the most critical phase of the landing operation, this was indeed a severe handicap for the supply units. But with a secret weapon worthy to be compared with the Trojan horse of the cunning Ulysses, the invasion armada solved the dilemma, bringing along their own – swimming – harbors. By "one brilliant technical idea," the colossal amounts spent on the Atlantic Wall, with its heavily armored harbor fortifications, had been rendered void, as Albert Speer, Hitler's master architect, admitted later in his memoirs.

Aerial photographs from the summer of 1944 show the Mulberry-idea in action. From the open invasion beach, four floating piers were reaching out – like the fingers of one hand – toward a teeming mass of landing boats. Around the improvised harbor basin, in a 4-mile semi-circle – "Phoenix" – caissons formed a jetty. Under cover of artificial breakwaters, Liberty Ships were unloading their cargoes, while barrage balloons were protecting Mulberry against enemy aircraft. In size, the whole installation was comparable to the harbor of Dover.

Arromanches, in the bay of which the prefabricated harbor had been kept, has since been given an honorary second name: Port Winston. It had been Winston Churchill's idea. In May 1942 he had already dictated an instruction for the Allied chief of operations, Lord Mountbatten, concerning "piers for operational use on beaches." "They must rise and fall with the tides. The problem of anchoring them must be solved. Have the best solutions worked out for me. Don't argue the matter; the difficulties will argue for themselves."

Just in time for D-day, about 15,000 British dock and building workers got the standardized concrete parts for two Mulberry units ready. The "monsters at sea," as they are described by the Ballantine pocket book on Allied Secret Weapons – The War of Science, were already installed according to plan in front of the beachheads, when, on June 19, 1944, a heavy storm rose up which was given the name Epsom. Just off the American Omaha-Beach, the worst northeasterly gale for forty years was literally playing football with the artificial harbor. Mulberry A had to be abandoned. Being better protected in the Bay of Arromanches, Mulberry B withstood the onslaught despite some bad damages. Though the construction had been planned for a working life of only three months, on the British Gold Beach supplies were still unloaded until December 1, 1944. Since then, the parts of the piers that had remained afloat have been swallowed by the sea or piled up on land. The last of the pontoons are looking out of boulders at the foot of the steep banks at Asnelles like moss-covered giant tortoises. At low tide one can reach a washed-up loading ramp on foot. The breakers are gradually grinding smooth the reinforced concrete of the platform: dead technology on the way to becoming megaliths. The unwieldy caissons, 230-foot-long huge blocks weighing 7,000 tons, are not likely to disappear from sight for a very long time yet. Winter storms have torn gaps in the jetty already. But the sea is not in a hurry; it might take a few centuries for the last breakwater to be dismantled.

West Side Manhattan/New York City

West Side Manhattan/New York City

West Side Manhattan/New York City

West Side Manhattan/New York City

West Side Manhattan/New York City

West Side Manhattan / New York City

West Pier, Brighton

West Pier/Brighton

West Pier/Brighton

West Pier / Brighton

West Pier/Brighton

Port Winston/Arromanches

Port Winston/Arromanches

Port Winston/Arromanches

NAVAL POWER OF YESTERYEAR
BRITAIN'S LAST AIRCRAFT CARRIERS

The former H.M.S. *Eagle* could complete her last voyage only with the aid of tugboats. The route led from the naval depot at Devonport on the English Channel round the Cornish peninsula toward Scotland. Heavy gales forced the convoy for two days to seek shelter off the coast of Ireland.

Several hundred spectators lined the banks, forming a farewell guard on honor, as the tug and tow were steaming up Loch Ryan. The rusty aircraft carrier did not just submit meekly to being ordered about. Shortly before reaching her goal, one of the hawsers tore, lashing out to port with the bows and gliding with a grating sound upon a sandbank. "Carrier *Eagle* fighting to the last," reported the local *Wigtown Free Press* under the dateline 26 October 1978.

With the next tide, the massive hull was got afloat again and towed to Cairnryan, a godforsaken nest on the west coast of Scotland. On the opposite side of the loch is Stranraer; the steam ferries to and from Ulster are passing daily in the haze. Cairnryan, too, has a harbor, a legacy from the last war. Later some of the Royal Navy's gray giants of steel could also be seen there, moored to the pier. Their traces were lost in the rainy desolation of Cairnryan; the harbor is used for the dismantling of old warships, the dead end of Britain's naval power of yesteryear.

In the course of World War II, the classical capital ships had lost their predominance upon the world's oceans. In the eyes of the English naval author Anthony Preston, the ship of the future, the more flexible aircraft carrier, "proved to be a hybrid creation both in its most simple and in its most sophisticated form, since it enabled man at the same time to conquer the air and the sea."

In the years of tribulation, the carriers of the Royal Navy had covered themselves with glory while hunting the German battleship *Bismarck* and during the bloody convoy battles to relieve Malta. In the inner sanctum of the Admiralty, plans were worked out for a mighty carrier fleet. To succeed the Grand Fleet it was to include more than twenty floating air bases; the plans provided for four units of 36,800 tons each and three of 45,000 tons each as the strategic backbone of Britain's sea power.

But as the empire fell apart, the new building program was more or less scrapped. Still, in the 1960s, Britain's surface navy included six new aircraft carriers, among them the 43,000-ton H.M.S. *Eagle* and H.M.S. *Ark Royal*, modernized for Buccaneer and Phantom jet planes.

Now, at Cairnryan, it was the end not only of the aged *Eagle*, the twenty-first warship carrying that name, but of Britain's large aircraft carriers generally. The demolition workers of Steel Supply Ltd. had hardly started on the 800-feet-long colossus, when in Devonport her younger sister ship put out her fires.

H.M.S. *Ark Royal*, flagship of the Royal Navy, had represented the British Crown at the bicentenary of the United States in 1976. Now, as was the *Eagle* in 1974, she was taken out of service without any replacement. The plans for a British supercarrier, the 60,000-ton *Furious*, had been shelved in 1966 for lack of financial resources.

The United Kingdom used the largest part of its naval budget for the expansion of a strategic underwater power which, with its computer technology and American atomic rockets, did not seem to fit in with the naval traditions of Trafalgar. "It was the end of an era for the Royal Navy, an era of marine flying, going as far back as 1912 and two former *Ark Royals*," wrote Anthony Preston in an epilogue on Britain's last fleet carriers.

Traditionalists and supporters of the Royal Navy refused to accept that the *Ark Royal* should be dismantled. They wanted to save her as a floating museum of Britain's role as a maritime power. But the costs of maintenance would soon have exceeded the financial resources of the donors, so that at the height of the protest movement a daily paper counseled a bit of realism: "Let the Old Lady die in Peace!" it said.

Meanwhile, at Cairnryan, one could witness the death of a 43,000-ton aircraft carrier. Starting with the tripod mast, in earlier times a typical feature of British warships, a team of forty men set about dismantling the *Eagle* deck by deck, with the welders to the fore. A private firm trading in steel aimed at winning in the course of two years' labor 30,000 tons of steel and 2,000 tons of nonferrous metal from the ship's hull, which after the disappearance of its superstructure and the flight decks had turned into an imposing offering cup.

With her military equipment removed, the former *Ark Royal* was waiting at Devonport for the highest bidder to turn her into scrap. In autumn 1980, when the tugs came to take her sad-looking mountain of steel to Cairnryan, the *Ark Royal* had already entered Britain's glorious pages of naval history. In a painting by naval artist Robert Taylor, also available as an inexpensive colored print, she is seen as queen of the Royal Navy, steaming full speed ahead – into the past. A mail-order firm was offering "a genuine piece of H.M.S. *Ark Royal* or H.M.S. *Eagle*." The relics – rum mugs, copper jars, ashtrays, ships' bells and similar marine knickknacks – bore the memorial emblems of the warships and were made out of original materials from the two carriers, according to an advertisement in the *Navy News*.

Aircraft Carrier Ark Royal/Cairnryan

Aircraft Carrier Ark Royal/Cairnryan

Aircraft Carrier *Eagle*/Cairnryan

Aircraft Carrier *Eagle*/Cairnryan

RUINS OF COAL AND STEEL
THE INDUSTRIAL EVOLUTION OF THE RUHR

"It is the most imposing panorama of industry, without equal. A wreath of seven ironworks on the banks of Ruhr, Rhine, and Emscher, a gallery of convincing greatness, interspersed with the pitheads of the collieries, the towers of the gasometers, and the white fumes from the coke-quenching towers, those huge billows of steam that vie with the clouds in the sky, presenting time and again the most beautiful spectacle in the district."

This is how Dr. Volkmar Muthesius in his book *Du und der Stahl* (You and Steel) sings the praises of the view from the bridge of the motorway Duisburg–Kaiserberg over the Ruhr District. That was in 1953. The "Ruhrpott," as it is called, had recovered from the havoc caused by the bombing raids during the war. In the skyline formed by mighty industrial plants, the writer saw reflected the vigor of a resurrected center of industry, reaching out as far as Dortmund toward "new goals."

The scenery of an area marked by a concentration of heavy industry is still the same in the Ruhr District, except that the once so imposing panorama of coal, iron, and steel is showing gaps and cracks. The set pieces of progress are crumbling apart. The old coal-mining district with its many pitheads and a forest of iron winding towers is a thing of the past. Now the *Manager Magazin* is speculating how the steel industry can get rid of its excessive and therefore costly working capacities, a problem that is facing the struggling steel trusts in other countries of the European community as well. Indeed, the industrial center spread out between Lorraine, Belgium, and Luxemburg can no longer hide the damages caused by the steel and coal crisis. Like remnants of an ancient culture of huge urns, one can see hot-blast stoves and blast furnaces rusting away side by side. The pressure of international competition forced the closure of ore mines and ore-dressing plants.

In the little steel town Rümelingen in Luxemburg, life circled for about a hundred years around the mining and smelting of the local Minette ore. It sounds like a divine judgment when the retired chairman of the regional mining inspectorate, Marcel Klein, says with regard to the end of the iron age in the town of the red rocks: "No one could believe that it would be coming, and suddenly it was there."

In 1956, no one in the Ruhr District thought that ten years later twelve million tons of the best coal would find no buyers and be piled up in huge pithead stocks around the miners' settlements. Mineral oil, swamping the German fuel market at prices that at today's reckoning must have seemed fabulously low, had replaced the coal, ruining the mining industry. With billions of marks, the federal government subsidized reorganization. Under the direction of the new coal combine, Ruhrkohle AG, the reorganization of the industry aimed at a completely mechanical extraction of the coal by the application of modern mining technologies, concentrating on only a few of the most profitable pits. This is why so many collieries had to be closed down.

The result in the district was a series of social shock waves. Almost 300,000 miners lost their jobs. Of the 158 pits that existed in 1956, only thirty were still being worked in 1980. Under the special conditions prevailing in the Ruhr District, once a pit is closed down it has to stay that way.

In the Muttenbach Valley south of Witten, somewhat outside the later industrial district, lies the cradle of Ruhr mining. Here, between slopes and meadows, small firms used to engage in open-cast mining, or driving small shafts straight into the mountain to get at the "black gold." For instance, in 1811 a team of five belonging to the union *Turteltaube* (Turtledove) produced 7,200 *ringel* of coal (67 tons).

The center of the district was the Muttenbach Valley house of prayer, where, until 1850, the colliers would meet for a common shift prayer. A bell in the louver would exhort the pitmen to be punctual; they sometimes came marching in from a good way away. The unique relic from the preindustrial age was preserved and is now classified as a historical monument. A historical path laid out for instructional purposes leads to further archaeological findings. It passes overgrown adits, old loading points, coal streets, and old colliery buildings. One can still see where, in 1829, a horse tram had been laid down. Mallet and crowbar, to this day the symbols of coal mining, were the standard tools. Today they are as rare as precious antiques, even for the museums hard to come by.

With the beginning of the machine age, coal mining spread out more and more to the lower lying seams north of the river Ruhr. The ground water was pumped up by steam power to allow for deep mining. The typical pitheads came into being, with their machine sheds, office buildings, and their boxlike coal-washing and coal-blending plants. Power stations and coking plants supplied the iron- and steelworks between Dortmund and Duisburg with unlimited fuel: steel in conjunction with the coal became the second leg upon which the heavy industry in the Ruhr District was based.

The writer Max von der Grün is not the only one to regard the Ruhrpott as "a capitalistic monstrosity"; now, under the shocks of a structural crisis, the district is again changing its physiognomy. Computerized and laser-directed mining robots are cutting through the bituminous coal at a depth of 4,000 feet. Another technology, in an experimental stage as yet, aims at finishing with traditional mining altogether and gasifying the coal at the source.

Anyway, with the last of the pitmen having been pushed into retirement, the pick and the pneumatic drill underground are finally passé. In those traditional mining towns such as Gelsenkirchen or Bochum, in the vicinity of which the coking plants and pit-shafts are lying idle, waiting to be pulled down or shut down, the Ruhr District presents a picture of an industrial antiquity. Bochum's last big colliery, Hannover, closed down in 1973. Seven years later the place looked like a field of ruins. The old Steag Power Station was half demolished already; the scrap was piling up along a siding ready for removal. The demolition division of the Thyssen-Sonnenberg Company had started on the remaining surface installations.

Between buildings of an indifferent functionalism, what struck the eye was the outline of a Malakoff tower – the name is derived from Fort Malakoff, which, during the Crimean War (1853–56), had offered most resistance during the battle around Sevastopol. In the Ruhr District, these brick towers with their imitation battlements indicated the transition from the rural mining of old to the large-scale mining underground on an industrial basis. Behind their fortlike exteriors, the Malakoff towers housed the headgear, including the capel wheel. Toward the end of the nineteenth century, they were gradually replaced by the skeleton constructions of the free-standing winding towers. At the "Hannover"-Colliery, the Malakoff tower is to be preserved as a relic of the past. Without it, the age of coal mining at Bochum would have been and gone without a visible trace.

Processing Plant, Audun-le-Tiche (ARBED)/Lorraine

Pit Entrance, Mine de Fer d'Angevillers/Lorraine

Steelworks Knutange (SOLLAC)/Lorraine

Steelworks Knutange (SOLLAC)/Lorraine

Steelworks Knutange (SOLLAC)/Lorraine

Backton Gasworks/East London

"Hannover" Colliery/Bochum

Coking Plant "Hugo" Colliery/Gelsenkirchen

Power Station, "Lorraine IV" Colliery/Bochum

Washroom, Ship Elevator, Henrichenburg/Westphalia

PYRAMIDS OF MODERN TIMES
GRAVEYARDS OF AUTOMOBILES
IN TOWN AND COUNTRY

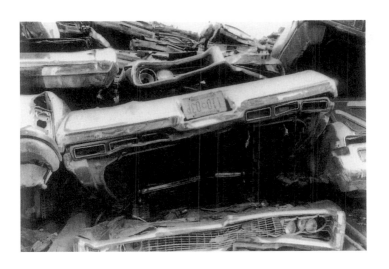

In the pyramids of the country of the Nile was expressed the veneration enjoyed by the Pharaohs. The huge tombs symbolized the central national cult in ancient Egypt: they were meant to secure beyond their death the absolute reign of the royal descendants of the sun god, the divine kings, as mediators between heaven and earth.

By comparison, the pyramids of wrecked cars express a central cult of the Western world, that is, the task to keep economic growth going by the production, sale, use, and scrapping of ever more automobiles.

In the United States, with 154 million cars the undisputed leader of the motorized world, there are monumental graveyards of cars. In Chicago, one single firm of car-breakers salvages 250,000 limousines annually. In Philadelphia, when you drive from the international airport across the Delaware toward the city center, you'll find on the right-hand side next to the bridge exit a stepped-up pyramid of thousands of flattened battleships of the road. Battered VW Beetles and other light-weight cars crown the scrapheap, which is the size of a three- to four-storied block of flats. A mountaineer might feel the urge to try his climbing skill on the steep walls, but the view from the top would hardly be rewarding.

In the scrapyard of Pollock & Abrams on Pennrose Avenue, about 140,000 cars annually are ground to scrap. The 4,000 h.p. hammer mill can manage an eight-cylinder car in one minute. Jarring tables, magnetic drums, and blowing engines separate the wheat from the chaff. The most important raw material to be salvaged for industrial recycling are fist-size lumps of up to 99 percent pure steel scrap. In the near harbor an Italian freighter is just waiting for the next load. That's why the heavy-duty trucks are on the go even on a Sunday. On each trip they carry the yield from about twenty car wrecks.

Local scrap merchants from Pennsylvania and the neighboring states supply the cannibalized car bodies. They go on a conveyor belt – minus the tires and the gas tank – and are crushed flat for better storage. Between 4,000 and 10,000 such wrecks are piling up on the pyramid according to the rate of supply and disposal. Pollock & Abrams buy the cannibalized consumer idols by weight. The current price is about $2.00 per hundredweight of vehicle, Manager Gary tells us. That amounts to roughly "sixty bucks" per discarded automobile.

Debris consisting of rubber and plastics, bits of bakelite, broken glass, spark plugs and similar small items is gathering around the foot of the dump: enduring artifacts that may one day lead some future archaeologist on the track of some fabulous motorman, whose disappearance coincided with a world fuel crisis coming dramatically to a head toward the end of the second millennium of the Christian Era. Rats and mice are quite obviously thriving in this oily sediment; in the upper tiers wild pigeons are coming to roost on this pile of a dead technology.

The late Erich Fromm, social philosopher and psychoanalyst, could see no contradiction in the adoration of the automobile and its shabby funeral. On the contrary: both, according to him, are part of the char-

acter of a consumers' society, in which the individual experiences himself and the world of merchandise no longer in terms of the utility value of a thing but in terms of the exchange value as defined by marketable features serving the consumer's image and prestige. That's why the affection one may have for one's current car is always short-lived, despite the proprietary pride it may engender ("you are what you own"). On this subject, Fromm wrote in his book To Have or to Be: "The car is no object that I am fond of, but a symbol of my status, my ego, an extension of my power. With the purchase of a car I obtain practically a new part-ego." There are many different motives, all of them constantly encouraged by car ads, for wanting to change one's car. Fromm refers to the increased pleasure gained by each new acquisition; the act of taking possession of a thing is a form of defloration. "Thirdly, every time you change cars you are offered a new chance to make a profit by the exchange, a desire deeply rooted in modern man. Yet another, fourth element is…the need for new attractions, as the old ones fade and become uninteresting after only a short time."

The car fetishism illustrates in an exemplary manner what Fromm in various books has described as the religion of the industrial age: a new "secret religion behind a Christian facade which reduces people to become the servants of the market and the machines which they built with their own hands."

Holy in this religion, according to Fromm, are labor, property, profit, and power.

The technical development of the car, at first regarded as a plaything of high society and the nouveaux riches, is due to a number of inventors. The American capitalist king Henry Ford practically invented the automobile a second time by presenting it to the industrial mass society as a golden calf. His spirit – the spirit of Detroit – hovers invisibly over the "new pyramids," for it wasn't until the democratization of the car that the economic dynamics of the internal combustion engine were set free.

With the announcement that he would produce a car for the masses, Ford shocked his competitors in 1909. His first mass-produced model was the "Tin Lizzie." Five years later, the gnarled car magnate set off the second shock wave to hit industry and financial establishment. On the 1st of January, 1914, he doubled the minimum wages of his workers from $2.50 to $5.00 daily, at the same time reducing the working time from nine to eight hours. That was ten times more than the average pay in the United States at that time. With the set purpose of making the blessings of the dawning motorized society available to the broad masses of workers and small farmers, Ford sold his Tin Lizzie as cheaply as possible. To this end, though, he radically increased productivity by the introduction of time-and-motion studies, resulting in a speedup of the conveyor belts, the splitting up of the working processes into single units, and the application of other measures making for maximum efficiency. Thus the price of a Tin Lizzie went down from the original $950 to only $365 in 1916. By 1927, when the legendary Model T had run its course, Ford had sold about 15.5 million cars.

More money = more buying power = more economic growth: the magical equation of American automobile capitalism had worked out. The workers at Ford's became car owners and the seemingly generous founder of the firm a multimillionaire. The nation becoming motorized was like the fulfillment of the American dream of unlimited wealth. Economic growth – as far as it was practicable – rolled from the assembly lines in the form of ever new luxury limousines.

The success story of a people's car in postwar West Germany marked the beginning of what became known as the "economic miracle." The old Volkswagen from Wolfsburg had already proved its mettle as a bucket car for Hitler's Blitzkrieg divisions. As a civilian car it even broke the production record of the Tin Lizzie. The way to a motorized society and, in conjunction with it, the growing Americanization of the Federal Republic were considered proof of progress in the heavy days of economic expansion in the 1950s and 1960s. With the cheap "Tiger in the Tank" (so went the ad slogan for gas in those years), the citizens of the Federal Republic were at last able to satisfy their urge to get about on four wheels. Upon the motorways, the rising standard of living showed itself by the increasing heraldry of the more powerful automobiles.

Temporarily, at least, the car euphoria quieted down under the impact of the oil crisis of 1973–74. But with the new boom in cars in 1976–78, the signs on the wall soon disappeared again. "With the car into the future" was the slogan of the International Automobile Association's motorcar fair in Frankfurt/Main, while huge traffic jams wrapped the city in a veil of smog.

The myth of the promised land still haunts predictions according to which the Federal Republic will draw roughly level with the United States in the year 2000, with regard to the number of cars and estate cars in relation to the population: 500 cars per 1,000 head. In 1979, the total number of private motor vehicles amounted to 22.5 million, or 368 units per 1,000 inhabitants, placing West Germany internationally in the middle of the statistical scale. This did not mean, according to the Federal Office for Motor Cars, that the market was saturated.

Even though the prestige of the motorcar may have suffered on account of the second oil shock and even more so because of the damages to the environment caused by the tin termites, as far as the builders of the new pyramids are concerned, they can cope with millions more. American-style shredding machines finish off wholesale any cars ready to be turned into scrap in the more densely populated areas. "When you get your new one, we'll do an expert job getting rid of your old one," says the advertisement of the "Autopresse," Tempelhof, in West Berlin's classified directory.

In the pseudorustic idyll of Upper Swabia, along the Federal Road 466, the car-processing firm of Georg Meyer is piling up "automobiles and trailers not in service and beyond repair." This is the official definition of a car wreck earmarked to be scrapped by the working group Removal of Refuse.

When the first great wave of cars landed in the scrapyards, the Board of Control issued instructions limiting the growth of the new pyramids. In a memorandum issued in 1976, it decreed detailed regulations as to the aesthetics of German car cemeteries. According to these, old cars or those that had been involved in accidents should be stored no longer than three months in stacks no higher than 10 feet. A salvage and scrapyard should cover an area not exceeding 43,000 square feet and the feeding area should be confined to a radius of 16 miles. A fence no less than 6 feet high was to enclose the piles of dead car bodies.

A concrete leftover like New York's ruined and traffic-wrecked, inner-city motorway along the Hudson is not so easily removed by decree. The bizarre monument to the god of motoring extends on pillars of steel from Forty-sixth Street right down to the southernmost point of Manhattan. It isn't worth repairing, but the chronic shortage of finances of the city of New York delayed its demolition as planned.

The six-lane West Side Highway was once, aside from the Long Island Expressway, the most frequented inner-city motorway, with an average of 130,000 cars daily going in both directions. For forty years the highway had been exposed to the heavy barrage of city traffic. In December 1973 it gave up the ghost of Detroit. Under the weight of a heavy-duty truck, the dilapidated concrete pavement collapsed and crashed to the ground.

"Soldiers! Four millennia are looking down on you!" Napoleon is said to have called out at the sight of the Pyramids of Gîzeh. We are not yet filled with such feelings of awe at the sight of this ruin of our civilization – worn down and corroded by stress and exhaust gases. For instance, on the seedy and run-down highway I once saw a dropout, one who had opted out of the automobile society. Instead, joggers, strollers, and cyclists from Greenwich Village had joined him. Paint-spraying artists turned the deserted roadway into a studio.

Automobile Scrapyard, Pollock & Aorams/Philadelphia

Automobile Scrapyard, Pollock & Abrams/Philadelphia

Automobile Scrapyard, Pollock & Abrams/Philadelphia

Car Utilization, Georg Meyer/Nattheim

West Side Highway/New York City

West Side Highway/New York City

DEATH IN TUCSON
THE LARGEST AIRCRAFT GRAVEYARD IN THE WORLD

America is letting its veteran air squadrons die in beauty. In Arizona, at the foot of the Blue Mountains, the superpower United States is maintaining the largest aircraft graveyard in the world. Approaching the Tucson International Airport for a landing, you can see it spread out to the south of the town, in the Sonora Desert, like a symmetrically patterned carpet. On a fenced-in piece of desert, "the third largest air force in the world" has been given the cocoon treatment.

The collection of several thousand laid-up flying machines in one place gives the surrounding tableland a touch of the unreal. In the changing interplay of light and shade, technology and nature merge into bizarre pictures. In the morning, there is a golden glimmer and glitter between the mile-long rows of aluminum bodies, as in a fairy tale landscape. In the heat of the midday sun, they fuse into glistening avalanches of metal, while at sunset, the multitude of aerodynamic outlines can be seen particularly clearly. Salvaged Stratofreighters loom like dying dinosaurs in the dusk – Death in Tucson.

In Tucson and its neighborhood, the patch of desert with its many aircraft corpses is known as the "boneyard." A closer look reveals that it is a depot for surplus or obsolete aircraft owned by the government of the United States. What the U.S. Air Force does not require between heaven and earth is taken to Tucson, to the Military Aircraft Storage and Disposition Center (MASDC).

The western desert of south Arizona is most suitable as an open-air hangar. In the dry, sunny climate, the aircraft need only a minimum of maintenance to escape corrosion. The crusty soil made of caliche, a mineral deposit, will do instead of concrete or steel parking ramps; even such heavy planes as the Boeing B-52 can be rolled to and fro on the plain ground.

At the end of World War II America had huge mountains of surplus war material. Between 1940 and 1945, after the mass production of arms of all sorts had been stepped up, the number of military aircraft to roll off the assembly lines amounted to 303,218, among them approximately 36,000 four-engined bombers. After the demobilization, the U.S. Air Force laid up part of its bomber and transport armada at the David-Monthan Base on the outskirts of Tucson. From 1946 to 1966 about 15,000 airplanes were put into mothballs at the MASDC. The majority of World War II aircraft have since disappeared in aluminum-smelting plants. MASDC bosses donated their last Boeing B-17 "Flying Fortress" to an aircraft museum in California over fifteen years ago. The only old-timers left at the turn of the year 1980–81 were the venerable C-47 Skytrain and her refurbished sister, the C-117 "Super Gooney," both military versions of the Douglas DC-3 Dakota. In a few years they are likely to have vanished from the graveyard scenery, just like the C-54 Skytrain: a four-engine Douglas transporter that was nicknamed "Raisin Bomber" during the Berlin airlift.

The jet age and the East-West conflict, in conjunction with Washington's generous military expenditures, made sure of new supplies. The predominant types of planes now aging away beneath Arizona's sun hail from the late 1950s and the 1960s. Frequently, they are what's left of the ten-year-long dance of death over Vietnam. In 1975, after the United States' military withdrawal from Indochina, the number of mothballed planes temporarily rose to about 6,200 jets, helicopters, and propeller-driven aircraft. The number has since been reduced to just about 4,000 machines – still enough for a gigantic air show on the ground.

There they stand, row upon row, as far as the eye can see: supersonic fighter planes, fighter-bombers, early warning aircraft, transport planes, trainers. The U.S. Navy and the Marine Corps have laid up enough carrier planes to form several squadrons; the U.S. Army left behind whole rows of helicopters and observation planes.

From afar, one can see shark fins rising from a dip on the ground: the tail units of about 200 B-52s standing close together. On approaching them one sees monstrous death birds, the so-called Stratofortresses. The camouflage paint is from Southeast Asia, where the intercontinental bombers, each with up to twenty-seven tons of high-explosive bombs in their loading bays, started off from Guam or Thailand to do their bit in the guerrilla war. Flying in at a height of 30,000 feet or more and, therefore, inaudible, they used to shed their loads over jungle forests and rice paddies, a form of aerial attack known as carpet bombing, which for those below must have seemed like the end of the world.

In addition to their war paint, the jet giants have been given a pop art disguise in Tucson, a white spray pattern. All parts on the fuselage that are movable, especially sensitive, or porous have been water- and sandproofed by a protective film of vinyl. The white skin over the glasswork reflects the sunlight, so that the interior of the fuselage never heats up to any extent, even when in summer the temperature exceeds 100° F. All explosives, weapons, secret equipment on board, and other special fittings have been removed from the B-52s for security reasons. The eight turbine units are lying greased in special containers underneath the wings. That's why the huge bombers cannot automatically be counted as "old iron." "Main Preservation July 1979" it says on the bow, meaning that they have again been given the standard conservation treatment for another waiting period of up to four years.

Embalmed, but technically intact, about half the MASDC fleet is just dozing away. The U.S. Defense Department is keeping these aircraft in reserve for various purposes. A certain proportion is returned to the operative units according to the changing military requirements. Also, Allied air forces, too, can draw on these reserves. It takes the MASDC maintenance staff about four weeks to bring any of these dead inmates of the depot to life again. In times of crisis, we were told, this could, of course, be speeded up.

Other types of planes are "demilitarized" for civilian use. The forestry department in California, for instance, took over reequipped patrol planes of the U.S. Navy as "firefighting bombers." Quite close by, on the border to Mexico, U.S. Customs agents are chasing dope smugglers in refitted Mohawk army scout planes. Some hundreds of sheriffs have changed from radio squad cars and trucks to ex-army helicopters, to supervise the traffic in their respective towns from the air.

However, most of the mothballed machines land sooner or later with their old comrades in the boneyard. In the rear part of the MASDC, which covers an area of approximately forty square miles, the rigid parade formation is increasingly falling apart. Here a ghost air force that will never take off again provides a bank of spare parts for the U.S. military. For years, the laid-up planes have been cannibalized systematically for all kinds of equipment.

In various stages of decay, they are waiting for their final end. The tires get slack, the paintwork is flaking off, the white plastic bandages peel off the fuselages in rags. Even in this desolate state, the 25-year-old Superconstellations still retain some of the grace that made them one of the most elegant aircraft toward the end of the propeller era. One is charmed by the harmony of their slender outline with the gently curved, trisected tail fin. But the traces of the long-legged, long-distance planes suffering from progressive deterioration are only too evident, even though their civilian sisters were still in service on many intercontinental routes in the 1960s. Some of these Superconnies have lost their radar noses; others had to yield up all four of their propellers or their engines. Showing no respect for the fading national emblems, whole colonies of pigeons have settled on the fuselages. In one case, their meeting place is a fuel tank in the wing; in another, they seem to prefer the humplike mounting of the radar.

A bit farther away, a formation of T-33 jet trainers appears to be getting ready for takeoff. The hoods of the cockpits are raised, as if awaiting the pilots; though these would have nothing but the mere shell of the two-seaters to get away in. The jet engines of the Lockheeds have been removed; the instrument panels have grown blind; all that is left of the electronic equipment are a few cable ends hanging loose.

The dilapidated Grumman E-2 Hawkeyes of the U.S. Navy seem to be bent under the weight of the plate-shaped, fiberglass radomes, where once the aerial of their omnidirectional radar was gyrating. Here, in alien surroundings, with their superimposed fittings, they look to us like giant wood lice dying motionless in the wilderness.

Among the most exotic inhabitants of this graveyard are a flotilla of Grumman Albatross hydroplanes that have got stranded in the desert. With their ungainly fuselages and their stubby landing gear, they are lying helplessly on the ground, the very examples of leaden ducks.

At dusk, the evening sun plays a spooky game with the scrap landscape. Just before it disappears below the horizon, it shines in some places through bull's-eyes and cockpits; but we get the impression that here and there the lights are going on in some of the cabins.

Big-bellied Boeing C-97s are looming into the air, as if they were their own tombstones. Some toppled over onto the tips of their noses; others toppled back onto their tails, as though they had no strength left to stand upright. Some of the invalids had to be jacked up or propped up with the aid of tubs to keep their balance. They've had their undercarriages amputated, or their wings, their tail units, or some other

parts. Their bowels are oozing out: rods, tanks, struts, lumps of electronics. With the continuing cannibalization, the superbirds gradually change into tin objects and fanciful sculptures. Fabulous creatures slip in and out of the skeletonlike frameworks, reminiscent of dragons, saurians, bats, or flying fish. Disarmament could, indeed, be a branch of art, but for the fact that we lack the political art of disarmament.

As these steel birds lose their functional forms, so the insignia and emblems of U.S. Icarus, too, suffer a transformation toward the surreal. Already, future archaeologists of aviation would have a lot of garbled numbers and word fragments to decode. On the nose of a battered jet bomber, the legend U.S. AIR FORCE has changed into U.S. AIR ORCE, for instance. On the damaged tail units of some Grumman Albatrosses, we spot some mysterious hieroglyphs. Only an expert would be able to interpret correctly RFOLK, SHINGTON or ANNAP as the home bases of the laid-up marine planes (NORFOLK, WASHINGTON, ANNAPOLIS).

The U.S. Air Force sees the MASDC in a different light, that is, in terms of its defense mission. In an information sheet, it praises the amassed salvage, worth about $5 billion, as the Pentagon's great savings account. Roughly 60,000–80,000 parts are returned to service annually, many of them no longer available elsewhere or extremely costly to obtain otherwise. Precious metals are recovered, too: for instance, gold wire from some electronic systems; there is platinum worth $2,500 in a single spark plug of the weighty Pratt and Whitney or Wright Cyclone star engines. All told, at MASDC, aircraft and parts recovered are worth about 25 percent of their original costs, according to the official statement.

What's left after everything has been salvaged that is worth salvaging gets the initials DPDO painted on the bow. It is the final death sentence. The resident defense property disposal officer has the task of selling to the highest bidder the residue waste material for melting down. An appropriate undertaker's business has established itself just outside the fence, complete with cranes and metal-cutting equipment. The business of South Western Alloys is flourishing. They are just salvaging the remainders of the T-29 Convair Liners. About 400 of these aircraft were used by the U.S. Air Force in the 1950s and 1960s as "flying classrooms" and liaison planes.

As on an assembly line leading nowhere, one after another of the twin-engined, propeller-driven planes are broken up. Combat aircraft, and in particular any strategic weapons systems, must be cut up and sold as scrap to "prevent them from falling into the wrong hands," according to the MASDC information sheet.

No fear that this could happen to the stripped atom-bombers in the far corner. Spread out on the ground, thirty-five B-52s are lying there in rows of five, as in a mortuary, slowly disintegrating into their component parts. The planks have been torn open; the fuselages are truncated; empty sockets stare out from the engine casings. Hares, wild rabbits, rattlesnakes, and other creatures of the desert have found a refuge between scraped shells of what not that long ago constituted the atomic sword of the American superpower.

From 1952–62 Boeing had supplied the U.S. Air Force with 774 Stratofortresses, costing $8 million each. In the critical days of the cold war, these intercontinental bombers, with the lightning emblem of the Strategic Air Command (SAC) under the cockpit, used to be in the air twenty-four hours a day. On their alarm patrols, they used to carry the atomic holocaust around half the globe. During a near catastrophe on January 16, 1966, four cylinder-shaped hydrogen bombs of 1.5-megaton explosive power each (by comparison: the Hiroshima bomb, 15 kilotons) were dropping from the sky over Spain's Mediterranean coast. A B-52 of the SAC and a jet fuel-supply plane had collided and exploded at a height of 33,000 feet.

Since 1975 these jet giants with a wingspan of 184 feet have been due for replacement. Compared with the apocalyptic perfection of their successors – AMSA (Advanced Manned Strategic Aircraft), a Mach 2.2, swing-wing aircraft with a nuclear bomb load of fifty tons – they are considered archaic. But since the B-1 bomber developed by Rockwell International is still in abeyance, the last of about 300 modernized B-52 Gs and Hs will remain until the end of the 1980s, if not until the end of the century, in the overkill arsenal of the Pentagon, as guided missile carriers.

Hence, the state funeral of the nightmare bomber might be delayed until the year 2000 or even later. That, at least, is what might be expected according to the current advance planning of the logistics computers in Washington. The question is whether the programmers think that far ahead.

Mortuary for B-52/Tucson

Lockheed Superconstellation with New Crew/Tucson

Torso of a C-47 (Dakota)/Tucson

  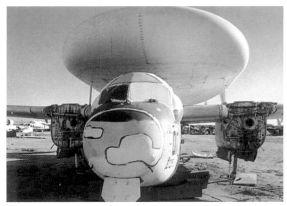

Graveyard Panorama/Tucson

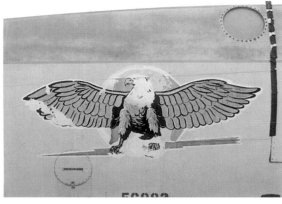  

Scrap Iconography/Tucson

Jet Skeletons/Tucson

Douglas C-47/Tucson

Grumman E-2 Hawkeye/Tucson

Graveyard Panorama/Tucson

Helicopter Insects/Tucson

EARLY SITES OF SPACE TRAVEL
CAPE CANAVERAL AFTER APOLLO

Mercury, Gemini, Apollo, Skylab – how long ago is that? At the cape where it all began, the past has begun already. The early sites of American space travel present themselves as crumbling ensembles of concrete that are reverting back to nature. Seemingly out of time, they are lying there without any purpose in the dune landscape of Cape Canaveral – like written-off stage props of a gigantic science fiction drama, of which no one knows when, how, and where it might end one day.

A lighthouse warden and a few farmers used to live on this isolated promontory on Florida's Atlantic Coast when, in 1947, the Defense Department in Washington acquired it as a restricted military zone.

The twenty-five square miles of sand bar, first inhabited by Indians, as can be seen by thousand-year-old burial mounds, was picked as the American counterpart to the former German military test station at Peenemünde on the Baltic Sea. A research team headed by Wernher von Braun had developed there, before and during World War II, the ballistic missile, A-4 or V-2. Leaving out of account its origin as a National Socialist armament project, it was a pioneering achievement that removed space travel from the realm of utopia to a technological possibility. In 1945, Wernher von Braun and the most important members of his staff entered the service of the Americans, who – just like the Soviets – were keenly interested in Hitler's last miracle weapon. An improved V-2, called Bumper, was the first rocket projectile to lumber from the testing ground at Cape Canaveral up into the sky. The date: July 24, 1950.

In the course of the Apollo boom, crowned at the end of the 1960s by two Americans landing on the moon, the remote coastal strip became an Eldorado of the space travel industry. The American National Aeronautics and Space Administration (NASA) established the J. F. Kennedy Space Center on Merritt Island, a peninsula close to the cape.

The space station on Cape Canaveral (for a time called Cape Kennedy) was a hive of activity day and night. At times, more than a dozen launcher rockets were assembled, tested, and finally tanked up for the countdown. Beside the Atlantic Ocean, one launching pad after another was lined up. With each new family of rockets, the assembly towers grew ever more imposingly into the sky – like a terrestrial staircase to the moon. Each of these assembly and launching places practically constituted an industrial complex known, in the NASA jargon, as a launch complex. Because the abandoned launch complexes were unsafe, the corroded service towers had to be blown up. The foundations and concrete structures are gradually covered by sand. Grass is growing on the underground control rooms. The spherical tanks for the liquid rocket fuel loom like mythological reliquaries into the world of the lagoons.

On the historical Complex 14 rises the astronomical symbol of the planet Mercury, in memory of the Mercury Program of 1961–63. A figure seven in a circle acts as a reminder of the first seven astronauts. At the base of the 5-foot-high stainless steel memorial are deposited, under a bronze plate, films, photographs, and scientific records about the various Mercury flights. These documents are meant to give people living in the third millennium an account of the history of the place. The founders of the memorial decreed that the strongboxes should not be unearthed and opened until the year 2464. They must have had in mind a solemn opening ceremony to celebrate the Mercury quincentenary, assuming, we must suppose, that in the meantime our spaceship earth would not suffer any lasting damage through our own space technology. At present, however, in the age of the space shuttles, killer satellites, and laser guns, the armament race of the superpowers is shifting ever more into space. With the world politically divided as it is, the possibility can no longer be excluded that, less than fifty years after Mercury and Wostok, man's departure to the stars might end with a cosmic Hiroshima: Cain and Abel as astronauts.

At Launch Complex 19, we come across a memorial tablet on which are inscribed the names of the Gemini pilots. Ten manned twin missions were started in 1965–66 where now the dead launching pad is standing. The launching platform, with the gripping devices for the 110-foot-high Titan II rockets, has been transmogrified into a mysterious steel gate. The umbilical tower with lightning conductor, red warning lamps, and supply systems lies tipped over on the two-storied concrete platform. The red lead oxide on the pylons shines in the rain; the panels are rattling in the wind. Abandon in Place is written on the Gemini torso, a millenary ruin that will surely outlast the space age of Cape Canaveral.

At Launch Complex 34, another memorial tablet reminds us of the hitherto greatest disaster in the U.S. space program. On January 27, 1967, a launch was simulated here for the first manned Apollo mission, when a fire broke out in the hermetically sealed command module. The crew of the spaceship, consisting of the three astronauts Grissom, White, and Chaffee, helplessly choked to death on top of their launch vehicle.

At the place of the tragedy rises the massive launch pad for the 230-foot-high Saturn IB. If we take a bird's-eye view of the circular plant, with what looks like an astro-altar in the center, the resemblance to a prehistoric religious site like Stonehenge, for instance, is evident.

"There will be no more one-way rockets like Saturn. Too much was lost," we were told by Dr. Kurt Debus, the most prominent old-timer from Peenemünde now at Cape Canaveral. Farther north, at Complex 39, the launching site of the manned moon expeditions, Debus erected a memorial to the mighty Saturn V. In contrast to the earlier space rockets, this three-stage supervehicle, with a 3,402-ton takeoff thrust, was not assembled and tested for months on the launching pad but in a huge hangar. After forty-five centuries, the Vertical Assembly Building (VAB) on Merritt Island replaced the Cheops Pyramid as the world's most voluminous edifice. With no comparable buildings close by, it is hard to realize that the cube-shaped colossus measuring 715 feet in length and 518 feet in width is approaching the height of the spire of Cologne Cathedral. All around it marshland is spreading, flat as an ironing board. Hidden behind the aluminum paneling is a 57,000-ton steel skeleton, resting on 4,225 steel pylons. Having a volume of 129,5 million cubic feet, the NASA cathedral could easily swallow the Pentagon, the United Nations skyscraper in New York, and then some. From the VAB, an electrically driven Caterpillar vehicle, called Crawler, brought the 365-foot-high Saturn V/Apollo combination to Launching Pad 39A at a snail's pace of 1 mile an hour. Including the rocket platform and the umbilical tower, the total load weight was 7,700 tons. The last time such a Saturn procession got going was on July 15, 1975, when a three-man crew of astronauts was launched to meet the Soviet spaceship Soyuz. As far as U.S. space travel is concerned, this marked the end of the one-way rockets in favor of the reusable Space Shuttle: a completely new launcher system, representing a mixture of rocket and gliding load carrier.

Dr. Debus, who as the first director of the space station had worked out and put into effect the mobile launching system, has retired, and Complex 39, with its assembling and launching facilities, is now the operational base for the Space Shuttle.

Near the VAB, where the fuel tanks for the Space Shuttle are set up, a Saturn V has been mounted as a key fossil of the Apollo era. The aluminum monster with a 33-foot diameter – they used to cost $248 million each – now serves as a backdrop for souvenir photographs. Sooner or later, it will fall victim to the corrosive influence of oxidation; the path from the technological-mathematical miracle to becoming a wreck is getting ever shorter.

Tourist buses are rolling through the J. F. Kennedy Space Center from eight in the morning until the sun goes down. There is a generously equipped lecture and exhibition center. NASA is presenting to the American taxpayer its space program, with all its blessings, in the form of a permanent public relations campaign. Disney World, Florida's major tourist attraction, is about an hour's car ride away, so that throughout the year there is never any shortage of visitors. In a souvenir shop with shelf space the size of a medium supermarket, space travel is commercialized on a retail basis. Here one can get everything from imitation spacesuits to cook books with the favorite dishes of the astronauts and a kitsch collection of beer mugs bearing the official NASA emblems of all the Apollo flights. Imitation moonstone made of plastic serves to decorate an astronautical writing desk; any number of miniature editions of the Space Shuttle were on sale even before its virgin flight: as pendant, tiepin, key fob, or bedside table lamp.

In what is known as the Rocket Garden, Wernher von Braun's scientific estate is put on show: seven different space projectiles that hit the headlines in the 1950s and 1960s. The points lashed tight, the bases in a concrete foundation, there they stand in a parklike landscape, staring upright into the sky. Space technology has long since outstripped these Juno, Redstone, Titan and Atlas rockets, or whatever else they were called. What is left is a phallocratic complex of pillars, without the warming and motherly roof.

Complex 34 (Apollo)/Cape Canaveral

Complex 34 (Apollo)/Cape Canaveral

Complex 34 (Apollo)/Cape Canaveral

Complex 34 (Apollo)/Cape Canaveral

Complex 34 (Apollo)/Cape Caraveral

Complex 19 (Gemini)/Cape Canaveral

Complex 19 (Gemini)/Cape Canaveral

Complex 19 (Gemini)/Cape Canaveral

Complex 19 (Gemini)/Cape Canaveral

Complex 19 (Gemini)/Cape Canaveral

Complex 19 (Gemini)/Cape Canaveral

"Rocket Garden"/Merritt Island

ROTTING PILES
THE RUINS OF THE ATOMIC STATE

"The power station Isar (KKI) was built in the Ohu District, on the left banks of the river Isar, about 9 miles downriver from Landshut in Lower Bavaria. In its immediate vicinity is the hydropower station Niederaichbach."

With its 427-foot-high chimney not to be overlooked, the nuclear power station Niederaichbach (KKN) also stands on a neighboring site. It doesn't quite fit into the picture of illustrated pamphlets (such as the one just quoted), when they proclaim nuclear energy as "the energy of the future." For the nuclear power station Niederaichbach, the future lies behind. It is Germany's oldest nuclear power ruin, though by no means the only one.

At Grundremmingen on the Danube, about twenty-five miles to the northeast of Ulm, the first commercially used atomic power station in the Federal Republic has passed away. Underneath its oval concrete casing, the 250-megawatt boiling-water reactor at Lingen on the river Ems is resting. All three reactors, all of them among the first in Germany, had closed down after less than fifteen years, although the electric power industry had assessed the working life of nuclear boilers to last thirty to forty years, generally.

Different from the conventional thermal power stations, nuclear reactors do not age only because of material fatigue and corrosion. Vital components such as pressure tanks, fuel jackets, or feed pipes subject to neutron irradiation grow gradually brittle. Former Federal Minister of Research Hans Matthöfer, in an interview he gave in 1977, asked whether the processes leading to embrittlement might not have been underestimated, that is that the presumable service life of nuclear power plants may have been overestimated.

The heavy water–pressure pipe reactor at Niederaichbach had no time left to turn brittle. It never worked properly, or, more correctly, it only worked for 18.3 days at full load. Siemens, the federal government and the Free State of Bavaria had erected the demonstration plant for the use of natural uranium as fuel in the meadows along the river Isar between 1966 and 1972, at a cost of 230 million marks, of which 140 million came from the taxpayers. After a test run of eighteen months, the "useless monster" (Die Zeit) was switched off without any further ado on July 31, 1974, to be taken out of commission altogether half a year later.

The fuel rods were returned to the nuclear research center at Karlsruhe; the heavy water worth 74 million marks – the brake fluid, so to speak, when atomic fission takes place – found a Japanese buyer.

What was left of the technical equipment and other fittings in the hot part of the power plant is now stored in the reactor building, "securely locked up." The other buildings of the defiant-looking concrete castle have been emptied; the set of turbines was dismantled and sold to a brown-coal-fired power station. They could be used as repositories, but nobody wanted to rent them.

A short circuit, with subsequent flooding of radioactive water in the 250-megawatt Block A at Grundremmingen, signaled the beginning of the end on January 7, 1977. An inspection revealed cracks in the boiler. In addition, after twelve years of service, a number of important parts of the safety layout such as the emergency cooling plant, the emergency controls, and the firefighting equipment did not pass inspection. The officially ordered refitting program finally added up to 250 million marks. The owners of the reactor were not willing to invest that much money on "pure chance," that is to say, without an official guarantee that they could continue operations on a long-term basis.

In 1977 too, the nuclear reactor at Lingen was closed down because of cracks in the steam transformer and defects in the emergency cooling aggregate. First there was talk of a thorough all-around overhaul, but it came to nothing because of the expected cost of 150 million marks.

On the other hand: neither at Niederaichbach, Lingen, or elsewhere can such obsolete atomic reactors be made simply to disappear. Unlike a written-off furnace, they cannot be got at with oxyacetylene blowpipes and pneumatic hammers. Rotting atomic piles go on being radioactive, the more so the longer the reactors had been in service. Scientists have found out that after three years of service, a 1,300-megawatt atomic reactor contains as much radioactivity as 4,000 Hiroshima bombs. From the active zone inside the reactor, known as the core, around which cooling agents are flowing, emanate radioactive particles and aerosols that enter the primary circuit. Because of the bombardment by neutrons, the pressure tanks, boilers, pumps, and pipe systems become radioactive. Other installations are contaminated by radioactive sediments right down to the foundations and the concrete shields. "If you enter them, you burn to death," warned the Sunday paper Welt am Sonntag, under the heading "How can we get rid of the atomic ruins?"

For the most dangerous radioactivity to abate, the ruins are left standing where they are. For ten, thirty, 100 years? The question has not yet been settled. In the Western industrial countries, the sum total of such ruins was thirty-eight in 1980, twenty-eight of them in the United States. By the year 2000, their number will double or even triple. In the Federal Republic, eleven more reactors will reach their technical age limit.

In the third millennium, the atomic graveyard will spread like a cancer. Two hundred twenty-four nuclear reactors, which in the middle of 1979 were operating with a total output of 124,586 megawatts, will then be heading for the final shutdown: seventy-two in the United States, twenty-eight in the USSR, and twenty-three in Japan.

What we may expect in the industrial conglomeration of western Europe can be gleaned from the Federal Parliament's publication 8-1977. It lists all nuclear reactors that are completed or in the course of construction in the nine countries of the European Community, as well as the probable decade within which they will have run their course. From 1981 until the year 2000, the number of reactors to be shut down is thirty-one; to these will be added another thirty-seven in the following ten years. From the year 2010 onward, yet another thirteen reactors built during 1981–84 would be due to be taken out of service.

Together with the eight nuclear ruins already existing, this would mean that in less than half a century the European Community could boast of eighty-nine ghost atomic power stations. No one today knows when and how to get rid of them safely. The vision suggests itself of a nuclear necropolis lasting a thousand years. "Such a building will last at least as long as the Egyptian pyramids," was the enthusiastic comment of the spokesman of a firm of reactor constructors.

According to the West German Federal Atomic Law, the owners of such reactors are obliged to remove the contaminated hulks. Practical experience in this respect is limited in Europe to the partial demolition of small test reactors like those at Lucens (Switzerland) and at Gross-welsheim (West Germany). The one exception so far was the United States, where at Elk River, Minnesota, a 58-megawatt nuclear reactor was completely dismantled in the years 1972–74.

At Niederaichbach, project director Hans Gallenberger intends to demonstrate how an industrial nuclear reactor can be razed to the ground without any danger to people or the environment. He announced that by 1985 the ground might be ready for Bavarian radishes to be planted there.

The demolition job (estimated cost: eighty million marks) that has thus been elevated to a pilot program is already in its planning stage. A specialist firm at Würzburg is developing welding equipment, cutting machines, clamping jaws, and metal saws to work by remote control, with which the contaminated reactor is to be dismantled and reduced to transportable pieces. Gallenberger is hoping to gain valuable know-how with regard to a particularly delicate problem in connection with the demolition of the reactor: the piecemeal, dust-free blasting of the biological shield, which at Niederaichbach has a thickness of 5 feet, but which is activated to a depth of 2 feet. The KKN will yield altogether about 1,200 tons of radioactive scrap and other dangerous materials that are to be gradually disassembled and placed in concrete-lined barrels for transport. To this must be added about 600–700 tons of activated concrete, out of a total of 130,000 tons.

What Gallenberger does not yet know is when he will be able to get rid of this radioactive waste. The problem of a final disposal at Gorleben or somewhere else is still in abeyance. "It is like sending Cologne Cathedral by parcel mail to an unknown address," mocked the magazine Stern in a report on the "State Funeral for 310 Million."

As compared with today's megawatt cathedrals, Niederaichbach, with an output of 100 megawatts, might be described as a mildly radioactive village chapel. In the case of a 1300-megawatt water pressure reactor, there would be approximately 6,500 tons of activated or contaminated steel installations alone to be dismantled, under quite different radiation hazards.

"With such large objects we lack the practical experience," is the conclusion of the safety expert at the Federal Ministry of the Interior, Wilhelm Sahl. This applies to the demolition technology as much as to the radiation protection for the staff and to the estimated costs. The

civil servant at Bonn was interviewed by the magazine Bild der Wissenschaft. His main theses: "The solution to most of the problems lies in the future ... Inquiries are far from being concluded.... There is still a long road ahead of us."

Both in industry and the official circles, there are but nebulous ideas about the costs involved in the shutting down and removing of the nuclear ruins. Sahl: "Opinions differ in that respect. According to estimates by American and German experts, about 20–30 percent of the original investments for a nuclear power plant; that would still amount to some 100 million marks." In a report of the U.S. Congress: somewhere between $31 million and $100 million. As the contempory encasing in concrete is as yet the cheapest method, Sahl is giving the engineers and scientists for the solution of all the problems involved, "at a rough estimate, half a century, based on the assumption that reactors like Biblis will be in service for thirty years – if there are no untoward incidents."

If .... For miles one can see the four vase-shaped cooling towers rising out of the valley of the Susquehanna River in Pennsylvania. From all directions they point the way to "the most expensive mausoleum in the world," as the contaminated nuclear center Three Mile Island (TMI) has been called since the near disaster at the Harrisburg reactor.

Day and hour literally burnt themselves into the memory of the people living in the nearby small town of Middleton: Wednesday, March 18, 1979, four o'clock in the morning. In Block TMI 2 (TMI 1 was being refueled), the 956-megawatt pressurized water reactor, working at a rated capacity of 98 percent, was getting out of control. The immediate cause: the cooling water supply system had broken down, due to the failure of two pumps.

But the fatal chain of human failings and technical hitches had already started – unnoticed – two weeks earlier, during maintenance work in the secondary circuit of the practically brand-new nuclear generator. The temperature and pressure in the 100-ton reactor core were rising far above the critical point. The 177 fuel elements filled with uranium were threatening to pour like burning lava through the steel containers and, on fusing, to vaporize into a poisonous cloud. "A bundle of immersion heaters in a kettle – and suddenly there is no water in the pot," was the way safety expert Sahl described the scenario of the threatening MCA (maximum credible accident). The raging uranium furnace switched itself off automatically, and the emergency cooling system pressed tons of water into the overheated pressure tanks, setting radioactive fumes free. The nuclear alarm was raised, and the population of Middleton fled in a panic.

Three Mile Island is now lying like a fossil on its island between the two arms of the river: like the grounded wreck of a ship with four disproportionate funnels. Tourists from all parts of the United States go there to gaze with amazement at this nuclear power mausoleum. To prove to those at home that they have actually visited Three Mile Island, they can bring back all manner of souvenirs, such as printed T-shirts or picture postcards showing the site "of the nation's worst nuclear accident." In inexpensive brochures, one can read the account of the "Days of Fear" by local chroniclers.

In protective suits and with breathing masks over their faces, a number of technicians inspected and filmed the contaminated reactor hall in the late autumn of 1980. The concrete floor was full of water stains and chemicals. On the videotape, one could make out the melted telephone. The decontamination and general cleanup of Three Mile Island will take seven years, at a cost of up to $1 billion, according to the estimates of the Edison Metropolitan Company. The damaged core of the reactor, still constituting a peril to the environment, cannot be approached to measure the radiation before the summer of 1985. Hence, it is uncertain for the time being whether the "Menetekel" of Harrisburg will one day be repaired, pulled down, or set in concrete for centuries to come.

Tugs and holiday steamers – we are back in the Old World – are chugging on the Danube past a deserted atomic town. There is not the least hint of steam circling out of the red-and-white painted chimney. At first glance, one cannot tell whether the power station has been switched off or never had been switched on. The signpost telling us the name of the place, Zwentendorf, and the little-used approach road provide the answer: before us lies "the only full-scale model in the world," the "Museum for Futile Technology," the "Bruno Kreisky Memorial" – all nicknames for the mothballed, 692-megawatt reactor near Vienna. Director Alfred Nentwich praised Austria's first nuclear reactor after it was completed as "an optimum of safety." This had its price: 1.2 billion marks, almost twice as much as originally estimated. Nevertheless, in a plebiscite for or against the reactor being put into service, on November 5, 1978, 50.48 percent voted against, pushing the uranium machine on to a dead siding. The Alpine Republic was almost split in half on the article of faith of whether or not atomic energy was needed for economic growth to continue. Now the political atomic ruin on the Danube, including its preserved entrails, is being kept intact for a new plebiscite at an annual maintenance cost of 13 million marks. The atom lobby and the founder of this monument, Bruno Kreisky himself, are relying on a change of opinion and on atomic energy from Zwentendorf from 1984 onward – provided no untoward incidents occur.

Atomic Power Station, Three Mile Island/Harrisburg

Atomic Power Station, Three Mile Island/Harrisburg

Atomic Power Station, Three Mile Island/Harrisburg

Atomic Power Station, Zwentendorf/Austria

Atomic Power Station, Niederaichbach/Bavaria

Atomic Power Station, Niederaichbach/Bavaria

BIBLIOGRAPHY

GENERAL INFORMATION
Michael Avi-Yonath / Emil Graeling: Die Bibel in ihrer Welt,
Konstanz 1960 (Fr. Rahn)
Kurt Benesch: Auf den Spuren großer Kulturen, Gütersloh 1979 (Lexikothek)
Warwick Bray / David Trump: A Dictionary of Archaeology, London 1970
(Penguin)
Erich von Däniken: Zurück zu den Sternen, Düsseldorf 1969 (Econ)
J. Sprague de Camp: The Ancient Engineers, New York 1974 (Ballantine)
J. G. Leithäuser: Die zweite Schöpfung der Welt, Berlin 1957 (Safari)
Ivar Lissner: Die Rätsel der großen Kulturen, München 1979 (DTV)
Artur Müller / Rolf Ammon: Die sieben Weltwunder, Bern u. München 1972
(Scherz)
Fritz H. Rienecker (Hrsg.): Lexikon zur Bibel, Wuppertal 1960 (Brockhaus)
Zeitschriften: Bild der Wissenschaft, Hobby, Manager Magazin, Der Spiegel

VERDUN / MAGINOT LINE / ATLANTIC WALL
Patric Boussel: Führer zu den Landungsstränden in der Normandie,
Paris 1974 (Presse de la Cité)
Philippe Carpentier: Arromanches et le débarquement allié du 6 juin 1944,
Arromanches 1975
Jean Compagnon: Les plages du débarquement, Rennes 1979 (France Ouest)
Remy Desquesnes: Le Mur de l'Atlantique, Bayeux 1976 (Editions Heimdal)
Keith Mallory und Arvid Ottar: Walls of War, London 1973
(The Architectural Press)
Albert Speer: Inside the Third Reich, New York 1970 (Macmillan)
Hans Speidel: Invasion 1944, Berlin 1974 (Ullstein)
Paul Virtlio: Bunker Archéologie, Paris 1975 (Centre Georges Pompidou)
Verdun: Illustrierter Führer durch die Schlachtfelder 1914–1918, Verdun o. J.
(Les Éditions Lorraines Frémont)
Christian Zentner (Hrsg.): Der Frankreich Feldzug 10. Mai 1940, Daten, Bilder,
Dokumente, Berlin 1980 (Ullstein)

»GRAVEYARD OF STEAM« AT BARRY DOCKS
Geoffrey Freeman Allen: Luxury Trains of the World, London 1979 (Bison Books)
Reg Cade (Ed.): Cade's Locomotive Guide, Bletchly (Marwain)
H. C. Casserley: British Steam Locomotives, London 1974 (Warne)
Brian Handly: Graveyard of Steam, London 1974 (Allen & Unwin)

SUNKEN HARBORS
Simon H. Adamson: Seaside Piers, London 1977 (Batsford)
Brian J. Ford: Allied Secret Weapons, New York 1971 (Ballantine)
Port Authority (Ed.): New York Port 1980–1981 Handbook, New York
West Pier Campaign: various brochures, Brighton

AIRCRAFT CARRIERS
Anthony Preston: Aircraft Carriers, London 1979 (Hamlyn)
Weyer: Taschenbuch der Kriegsflotten, München (J. F. Lehmann's)

COAL AND STEEL
Wilhelm Dege: Das Ruhrgebiet, Braunschweig 1972 (Vieweg)
8. Europäischer Knappentag: Festbroschüre, Rümelingen 1979
Manager Magazin 6/1981: Stahlindustrie: Nur noch Schrott?
Merian 8/33: Ruhrgebiet, Hamburg 1980 (Hoffmann & Campe)
Volkmar Muthesius: Du und der Stahl, Berlin 1953 (Ullstein)
Gustav Adolf Wüstenfeld: Frühe Stätten des Ruhrbergbaus, Wetter 1975

GRAVEYARDS OF AUTOMOBILES
Michael Busse: Autodämmerung, Frankfurt 1980 (Fischer)
Erich Fromm: To Have or to Be?, New York 1976 (Harper & Row)
Heinz Sponsel: Henry Ford, Gütersloh 1960 (Mohn)
Helmut Swobodo: Der Kampf gegen die Zukunft, Frankfurt 1980 (Fischer)

AIRCRAFT GRAVEYARD TUCSON
Peter M. Bowers: Boeing B-52 A/H Stratofortress, London 1972 (Profile)
William Green & Gordon Swanborough: Military Aircraft Directory,
London 1975 (Warne & Co)
Karlheinz Kens: Die alliierten Luftstreitkräfte 1939–1945,
München 1962 (Moewig)
Military Publishers: Davis-Monthan AFB The Old Pueblo, San Diego 1980
Dietrich Seidl: Flugzeuge '77, Stuttgart 1977 (Motorbuch)
U.S. Air Force Fact Sheet: MASDC – The Military's Moneysaver,
Davis-Monthan AFB 1979

CAPE CANAVERAL
George Alexander: Moonport USA, J. F. Kennedy Space Center 1977
Michael Maegrith: Mondlandung, Stuttgart 1977 (Belser)
Otto Merk: Raumfahrt Report, München 1967 (Bruckmann)
NASA: Facts Book and various brochures, J. F. Kennedy Space Center
U.S. Air Force: From Sand to Moondust, Cape Canaveral

RUINS OF THE ATOMIC STATE
Bild der Wissenschaft (Dez. 1979): Wohin mit den alten Kernkraftwerken?
Robert Jungk: Der Atom-Staat, Reinbek 1980 (Rowohlt)
Kernkraftwerk Isar (Hrsg.): Kernkraftwerk Isar, 1979
RWE (Hrsg.): Mehr Wissen über Strom und Kernkraft Nr. 14, Essen
Der Spiegel (Nr. 15/1979): Alptraum Atomkraft
John C. Staley / Roger R. Seip: A Time of Fear, Harrisburg 1979
Holger Strohm: Friedlich in die Katastrophe, Frankfurt/M. 1981 (Zweitausendeins)
Anton Zischka: Kampf ums Überleben, Düsseldorf 1979 (Econ)

MANFRED HAMM
Born at Zwickau, Germany, in 1944, Manfred Hamm lives in West Berlin and Paris as a photographer, working for German and foreign magazines and art publishers. His picture book *Berlin — Landscapes of a City* was published in 1977. For each of his picture books published by Nicolai, *Berlin — Monuments of an Industrial Landscape* (1978) and *Cafés* (1979), he was awarded the Kodak Photo Book Prize.

ROBERT JUNGK
Born in 1913, journalist and futurist, Robert Jungk became internationally known after World War II for his collection of documentary essays *The Future Has Already Begun*. Pointing out the dangers of unleashed atomic power, he wrote his Hiroshima report *Radiation from the Ashes* and *Nuclear State*. He lives in Salzburg, Austria, and holds a lectureship in futurology at the Technical University, West Berlin.

ROLF STEINBERG
Born in 1929, Rolf Steinberg lives in Berlin as a freelance journalist and book editor. Having studied political science, history, and journalism, he has worked as a reporter for international news agencies, Paris correspondent for the West German news magazine *Der Spiegel,* and as a publisher's reader.